Mohamed Ridda Laouar ·
Valentina Emilia Balas · Vincenzo Piuri ·
Dana Rad · Zineb Touati Hamad ·
Abbas Cheddad
Editors

13th International Conference on Information Systems and Advanced Technologies "ICISAT 2023"

New Trends in Artificial Intelligence, Computing and Decision Making. Volume 1

Springer

Editors
Mohamed Ridda Laouar
LAMIS Laboratory
Université Larbi Tébessi
Tebessa, Algeria

Valentina Emilia Balas
Faculty of Engineering
"Aurel Vlaicu" University of Arad
Arad, Arad, Romania

Vincenzo Piuri
Department of Computer Science
University of Milan
Milan, Milano, Italy

Dana Rad
Psychology and Social Sciences
Aurel Vlaicu University of Arad
Arad, Romania

Zineb Touati Hamad
LAMIS Laboratory
University of Tebessa
Tebessa, Algeria

Abbas Cheddad
DIDA
Blekinge Institute of Technology
Karlskrona, Sweden

ISSN 2367-3370 ISSN 2367-3389 (electronic)
Lecture Notes in Networks and Systems
ISBN 978-3-031-60590-1 ISBN 978-3-031-60591-8 (eBook)
https://doi.org/10.1007/978-3-031-60591-8

This Springer imprint is published by the registered company Springer Nature Switzerland AG
The registered company address is: Gewerbestrasse 11, 6330 Cham, Switzerland

Lecture Notes in Networks and Systems 981

The series "Lecture Notes in Networks and Systems" publishes the latest developments in Networks and Systems—quickly, informally and with high quality. Original research reported in proceedings and post-proceedings represents the core of LNNS.

Volumes published in LNNS embrace all aspects and subfields of, as well as new challenges in, Networks and Systems.

The series contains proceedings and edited volumes in systems and networks, spanning the areas of Cyber-Physical Systems, Autonomous Systems, Sensor Networks, Control Systems, Energy Systems, Automotive Systems, Biological Systems, Vehicular Networking and Connected Vehicles, Aerospace Systems, Automation, Manufacturing, Smart Grids, Nonlinear Systems, Power Systems, Robotics, Social Systems, Economic Systems and other. Of particular value to both the contributors and the readership are the short publication timeframe and the world-wide distribution and exposure which enable both a wide and rapid dissemination of research output.

The series covers the theory, applications, and perspectives on the state of the art and future developments relevant to systems and networks, decision making, control, complex processes and related areas, as embedded in the fields of interdisciplinary and applied sciences, engineering, computer science, physics, economics, social, and life sciences, as well as the paradigms and methodologies behind them.

Indexed by SCOPUS, EI Compendex, INSPEC, WTI Frankfurt eG, zbMATH, SCImago.

All books published in the series are submitted for consideration in Web of Science.

For proposals from Asia please contact Aninda Bose (aninda.bose@springer.com).

Contents

Bridging Sky to Network: FSO in 5G

Oumaima Allaoua[1], Hanane Djellab[2(✉)], Riad Saidi[3], Bentahar Tarek[4], Fouzia Maamri[5], and Ahmed Mohamed Salem[1]

[1] LAMIS Laboratory, Department of Electrical Engineering, Larbi Tebessi University, Tebessa, Algeria

[2] LTI Laboratory of Guelma, Department of Electrical Engineering, University Larbi Tebessi, Tebessa, Algeria

hanane.djellab@univ-tebessa.dz

[3] Department Commun Base Science and Technology, Mustapha Ben Boulaid Batna2 University, Batna, Algeria

[4] Department of Electrical Engineering, University Echahid Larbi Tebessi, Tebessa, Algeria

[5] SATIT Laboratory, Department of Industrial Engineering, University of Abbes Laghrour, Khenchela, Algeria

Abstract. A new optical transmission technique that offers high data speeds is known as free space optics (FSO), wide bandwidth, free licensing, and strong security. It presents an attractive solution to meet the connectivity needs of 5G, which has revolutionized our lives, as well as the requirements of the Internet of Things (IoT). However, FSO communication faces challenges, particularly in adverse weather conditions. This paper focuses on analyzing the performance of FSO links based on Orthogonal Frequency Division Multiplexing (OFDM) for 5G and IoT applications. Our study aims to evaluate the effectiveness of FSO technology by considering key factors such as data rate, laser power, bandwidth, and range. The obtained results demonstrate the successful achievement of a data rate of 10 Gbit/s, a laser power of 10 dBm, and a wide bandwidth of 6 GHz, which are suitable for 5G applications. Moreover, under clear weather conditions, a maximum range of 10 km has been achieved. The performance of the system is seriously affected by weather conditions such as snow and rain, and this must be noted in order to highlight the need for more focus in real FSO installations for 5G and IoT networks.

Keywords: FSO · 5G · optical network · free space optical · EVM · constellation · Link

1 Introduction

Rapid changes are being made to the telecommunications industry to keep up with users' increasing desire for constant access, regardless of where they are. As people look for new applications and services, technological development and societal needs have been the driving drivers behind this growth. Starting with the early intelligent telephone networks and continuing with today's high-speed vast area networks, communication networks have experienced constant development to keep up with these changing needs.

M. R. Laouar et al. (Eds.): ICISAT 2023, LNNS 981, pp. 1–10, 2024.
https://doi.org/10.1007/978-3-031-60591-8_1

However, the issues brought on by the rise in smartphone users and the extensive usage of social networking applications have not been entirely resolved by the introduction of fifth-generation (5G) networks. Due to the exponential rise in data consumption, the telecoms industry is currently experiencing a bottleneck as the demand for high-speed data rates continues to make it up. To offer smooth connectivity to users, this bottleneck presents a considerable barrier that needs to be overcome [1].

A type of optical-wireless communication called free-space optical communication (FSO) has surpassed conventional radio frequency (RF) communications in popularity [2, 3]. FSO communication has a number of benefits, including fast data transfer, secure communication, no use of electromagnetic waves, and no licensing requirements [4].

The ability to deploy FSO communication in a variety of locations and use the atmosphere as a medium for information signal transmission between the transmitter and receiver is another benefit. For best effectiveness, it does need a line of sight (LOS) link [5, 6]. In order to meet the bandwidth needs of data-intensive applications like social networking sites, video conferencing, high-definition television, cloud computing, and over-the-top services like Amazon Prime and Netflix, FSO networks offer high-bandwidth wireless communication links [5, 6].

Considering all of the benefits of FSO communication, it can be successfully used in new applications that need for fast transmission speeds. For instance, it can be used for 100 Gbps or more data transmission between high-altitude platforms and optical ground stations. Similar to how fixed-wing aircraft and hospitals may communicate using FSO technology to transmit data at speeds up to 1 Gbps. Furthermore, as shown in Fig. 1, it can satisfy the data communication requirements between quadcopter drones and residential buildings.

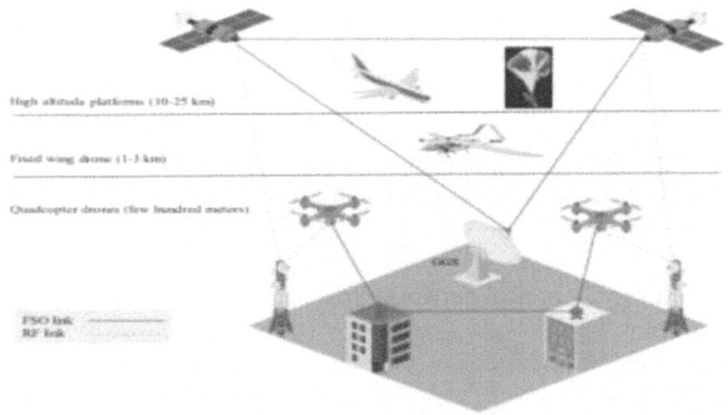

Fig. 1. FSO links have diverse applications in 5G and beyond networks [7].

However, diverse weather elements including snow, fog, and dust cause optical beam attenuation, which degrades the signal and affects the overall performance [8, 9]. The paper is structured as follows: Sect. 2 provides an overview of FSO technology, Sect. 3 presents the FSO system, Sect. 4 discusses the results obtained, and finally, Sect. 5 presents the conclusion.

2 FSO Technology

An advanced method called free-space optics (FSO), also known as fiber-free or fibreless photonics, permits the transmission of modulated light pulses by the free space medium, such as air or the atmosphere, to achieve high-speed and broadband communications [10]. In FSO communication, digital data is transmitted across the free space atmosphere using a narrow, tightly focused, and modulated optical laser beam, which is then received and processed at the receiving station [11]. It should be noted that line-of-sight (LOS) is a key component of FSO communication.

This means that at each networking location, the FSO transmitter and the FSO receiver must have a clear, straight line of sight in order to successfully communicate. The LOS condition makes sure that there are no natural obstructions, like trees or buildings, in the way of the transmission path between the two ends. This requirement is illustrated in Fig. 2, which demonstrates how a clear line of sight between the FSO transmitter and the FSO receiver is necessary for optimum performance. FSO lines show potential uses for 5G and future networks in a number of different applications. To satisfy the growing demand for data-intensive applications in urban settings, FSO lines provide a stable and fast connectivity option.

In addition, they offer a practical last-mile connectivity choice, bringing high-bandwidth services to underserved regions. As they efficiently manage the data transfer between cellular base stations and core networks, FSO connections are perfect for mobile backhaul. FSO lines are essential for quickly restoring communication infrastructure in disaster recovery situations. FSO links also make it possible for IoT devices to connect securely and effectively, facilitating the growth of IoT applications. The large capacity and secure communication channels of FSO networks are useful to the military and defense industries.

FSO links are also used in remote sensing and monitoring, which makes it easier to monitor and keep an eye on the environment. In general, FSO connections offer adaptable solutions across several domains to satisfy the evolving demands of 5G and future networks.

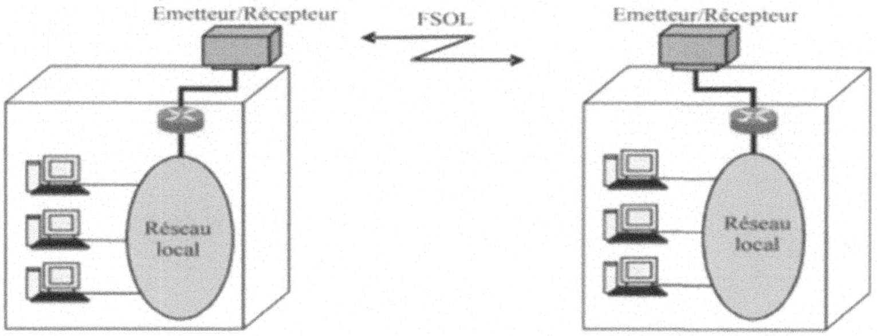

Fig. 2. The transmission in Free Space Optics (FSO) communication [12]

3 FSO System Description

The FSO system is composed of interconnected components that facilitate the transmission and reception of optical signals, as depicted in Fig. 3. The signal transmission process begins with a Pseudo Random Binary Sequence (PRBS) generator, which generates a 16-level Quadrature Amplitude Modulation (QAM) signal. This QAM signal is then fed into an Orthogonal Frequency Division Multiplexing (OFDM) modulator, which converts the serial data into parallel data using orthogonal multiplexing techniques. To optimize the bandwidth of the OFDM signal's subcarriers in the time domain, a Low-Pass Cosine Roll-Off filter is applied.

This filter is separately applied to the in-phase (I) and quadrature (Q) components of the OFDM signal, operating at a radio frequency of approximately 6 GHz.

Its purpose is to reduce the bandwidth and mitigate interference between subcarriers. The modulated data is then routed to a quadrature modulator, which performs the final modulation step before transmission. The modulated signal is used to modulate a Continuous Wave (CW) laser using a Mach-Zehnder (MZ) modulator. This process results in a modulated optical signal that is transmitted through free space, with the aid of amplifiers to enhance the power of the FSO signal.

On the receiving end, a PIN photo detector is employed to receive the optical signal and convert it into an electrical signal. To extract the original data, an OFDM demodulator and a QAM demodulator are utilized. These components reverse the modulation and multiplexing steps performed at the transmitter, effectively decoding the received signal.
On the receiving end, a PIN photo detector is employed to receive the optical signal and convert it into an electrical signal. To extract the original data, an OFDM demodulator and a QAM demodulator are utilized. These components reverse the modulation and multiplexing steps performed at the transmitter, effectively decoding the received signal.

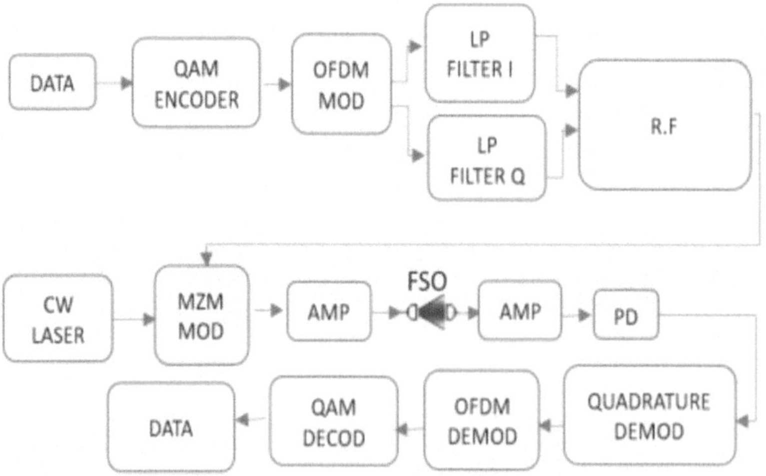

Fig. 3. Diagram illustrating the FSO system configuration.

Lastly, a constellation viewer is utilized to analyze and visualize the system's output. This component allows for the assessment of transmission quality and accuracy, providing valuable insights into the performance of the FSO system.

4 Results and Discution

The system's performance is assessed by considering various parameters, including the data rate, communication distance, and the impact of meteorological conditions. The Error Vector Magnitude (EVM) is used as a performance metric to evaluate the accuracy and quality of transmission. The specific parameter values used in the evaluation are provided in Table 1, reflecting practical models.In the transmitter section, the constellation viewer is employed to observe the generated signal's constellation (refer to Fig. 4) and obtain an ideal constellation where the points are well-defined and evenly spaced. This ideal constellation serves as a reference for calculating the EVM. Constellation analysis and EVM calculation are crucial tools for assessing transmission quality in our free-space optical communication system. They allow us to measure the deviation between the actual positions of the constellation points and the ideal positions, enabling us to quantify the precision and fidelity of the transmitted signal.This following table shows the parameters used in the simulation.

Table 1. Simulation's parameters

Parameter	Value
Data rate	10 Gbit
Power of spatial laser	10 dBm
Laser Wavelength	193.1 THz
Radio frequency	6 GHz
Sequance length	262144 Bits
Transmitter aperture diameter	5 cm
Receiver aperture diameter	20 cm
Beam divergance angle	2 mrad
FSO range	From 200 m to 10 km

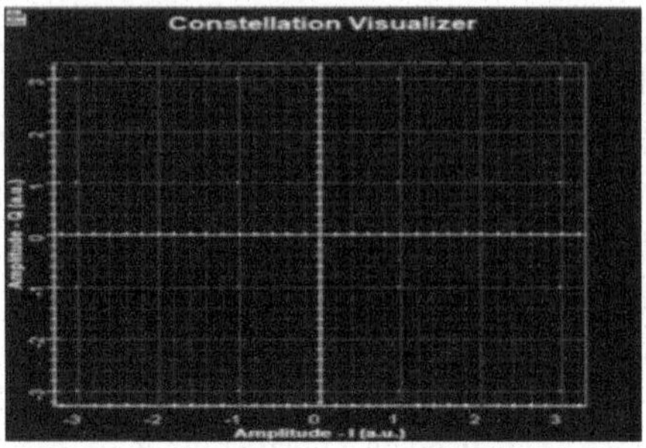

Fig. 4. Constellation in the transmitting section

4.1 FSO Range

To evaluate the performance of our system at different distances, we conducted experiments with a fixed data rate of 10 Gbps and an optical source power of 10 dBm in clear weather conditions. We systematically varied the distance from 200 m to 10 km and analyzed the resulting transmission quality using the Error Vector magnitude (EVM) metric. The EVM values obtained for each distance are illustrated in Fig. 5, providing insights into the relationship between distance and transmission quality in our FSO system.

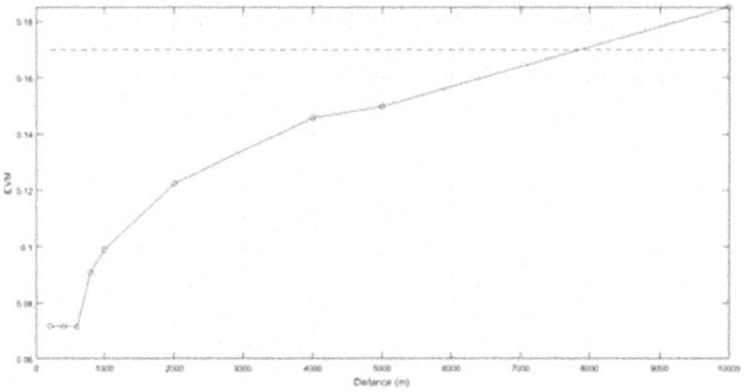

Fig. 5. EVM Variation with Distance

As the distance increases, the EVM exhibits a noticeable rise, indicating a degradation in the link quality beyond a threshold of 0.17%. This observation highlights the impact of distance on the fidelity of the transmitted signal.

Figure 6 displays the constellations obtained at distances of 10 km, allowing us to visualize the effects of increasing distance on the signal fidelity and identify any distortions that may occur.

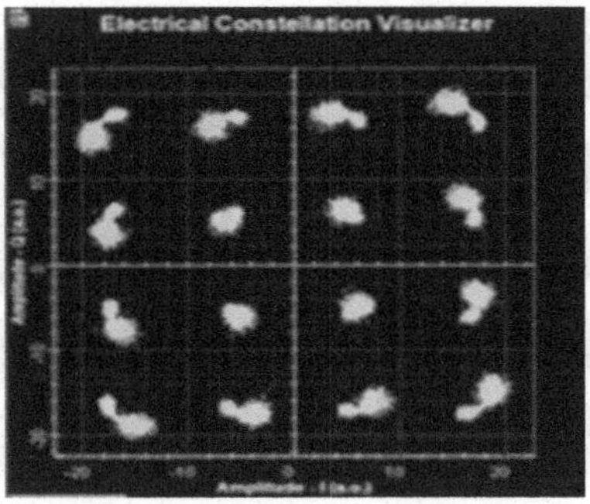

Fig. 6. Constellation Comparaison: 10 km Distance Variation

4.2 Effect of Data Rate on the FSO Link

In this section, we investigate the impact of modifying the data rate generated by the PRBS (Pseudo-Random Binary Sequence) on our system, building upon the findings from the previous distance-variation analysis. We maintain a fixed link distance of 600 m and vary the data rate within the range of 1 Gbps to 40 Gbps. By examining the resulting constellations at different data rates (6G, 10G, 20G, 40G) as illustrated in Fig. 8, we can assess the effect on transmission quality.

According to the results depicted in Fig. 7, as the data rate surpasses 30 Gbps, the link quality starts to deteriorate, resulting in an EVM increase of more than 0.17. This suggests that pushing the data rate beyond this threshold adversely affects the overall performance and fidelity of the transmitted signal.

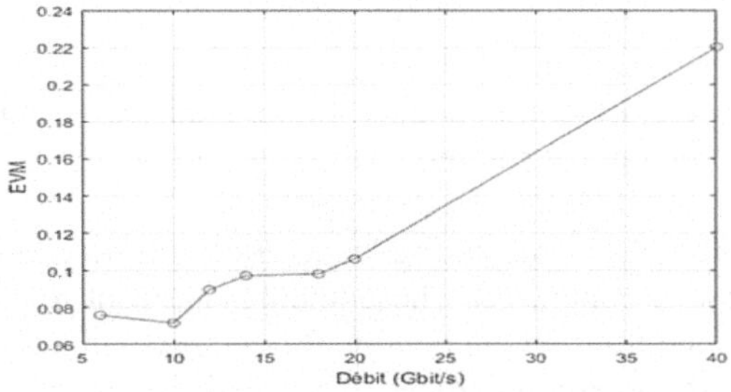

Fig. 7. EVM versus da ta rate

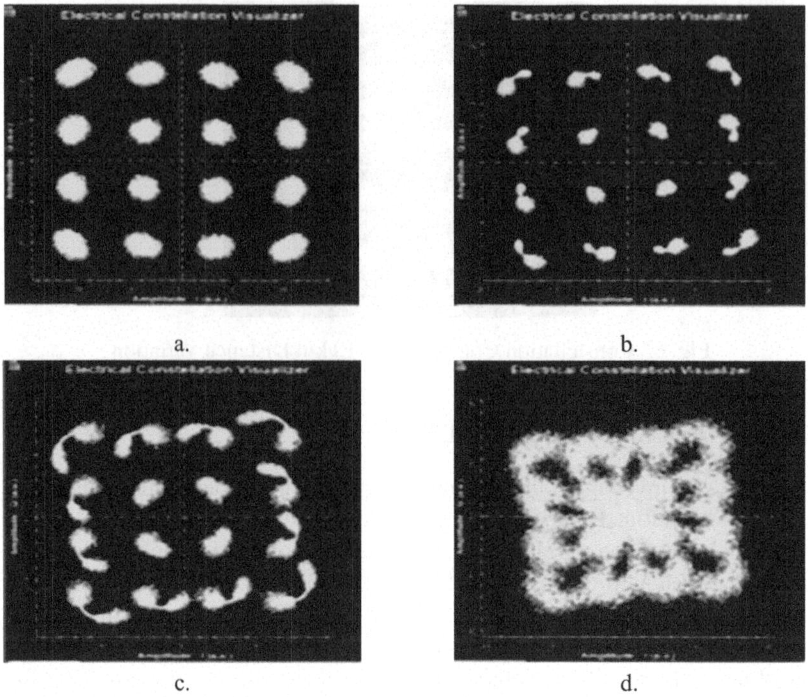

Fig. 8. Constellations at Varied Data Rates, a. constellation for 6 Gbit b. constellation for 10 Gbit, c. constellation for 20 Gbit, d. constellation for 40 Gbit

4.3 Effect Meteorological Condition on the FSO Link

Weather conditions pose a significant challenge for free-space optical links, as they act as obstacles and adversely affect link quality. In our study, we examined the impact of weather conditions on our optical communication system, focusing on the EVM metric. We maintained a data rate of 10 Gbps for our analysis.

Figure 9 presents the results, depicting the relationship between the EVM and distance for different weather conditions. These findings allow us to gain insights into the influence of weather conditions on the quality of free-space optical transmission.

The results of our study clearly indicate that weather conditions have a significant impact on the performance of free-space optical links. Adverse weather conditions, such as heavy rain or dry snow, act as obstacles in the path of the optical signal, leading to degradation in link quality.

In contrast, under clear weather conditions, the link demonstrates optimal performance, with the ability to achieve longer distances while maintaining a desirable EVM below 0.17.

These observations underscore the importance of considering weather conditions in the planning and deployment of free-space optical communication systems. By accounting for these conditions, engineers and network planners can optimize link performance and ensure reliable transmission quality in various weather scenarios.

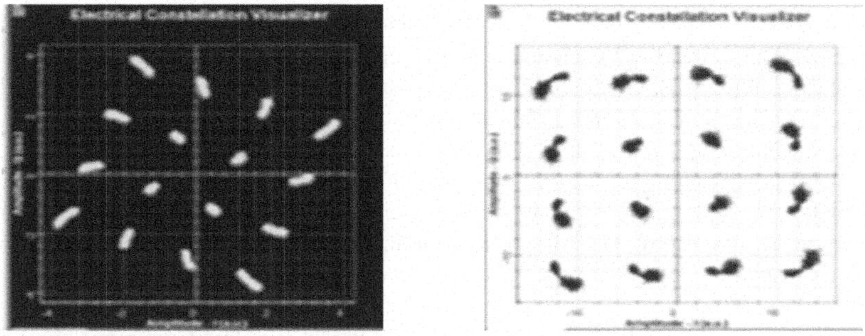

a. b.

Fig. 9. Constellations for different weather conditions and distances, a.constellation for heavy rain for 4 km, b.constellation for distance 600 m for clear weather

5 Conclusion

Our article focuses on evaluating the performance of an open-space optical communication system using FSO technology in conjunction with grows of 5G. We then ran a simulation using an OFDM modulation and a 10 dBm laser to achieve high-bandwidth 6 GHz connectivity, 10 Gbit/s speed over a distance approximately to 10 km, and an EVM of less than 0.17 under clear weather conditions. However, we have also found that unfavorable weather conditions, such as heavy rainfall or dry snow, can result in a significant reduction in the transmission distance.

In conclusion our article contributes to the advancement of knowledge in the field of optical communications in free space and provides valuable information for the implementation of effective solutions in the context of 5G next generation mobile networks.

It also paves the way for future research and development to improve the performance and reliability of optical communication systems in open space for new generations of mobile communication.

References

1. El-Mottaleb, S.A.A., Singh, M., Chehri, A., Ahmed, H.Y., Zeghid, M., Khan, A.N.: Capacity enhancement for free space optics transmission system using orbital angular momentum optical code division multiple access in 5G and beyond networks. Energies **15**, 7100 (2022)
2. Dat, P.T., Kanno, A., Yamamoto, N., Kawanishi, T.: Seamless convergence of fiber and wireless systems for 5G and beyond networks. J. Light. Technol. **37**, 592–605 (2018)
3. Kanno, A., Inagaki, K., Morohashi, I., Sakamoto, T., Kuri, T., Hosako, I., Kawanishi, T., Yoshida, Y., Kitayama, K.: 40 Gb/s Wband (75–110 GHz) 16-QAM radio-over-fiber signal generation and its wireless transmission. Opt. Express **19**, 56–63 (2011)
4. Mostafa, S., Mohamed, A.E.-N.A., El-Samie, F.E.A., Rashed, A.N.Z.: Performance evaluation of SAC-OCDMA system in free space optics and optical fiber system based on different types of codes. Wirel. Pers. Commun. **96**, 2843–2861 (2017)

5. Pottoo, S.N., Goyal, R., Gupta, A.: Performance investigation of optical communication system using FSO and OWC channel. In: Proceedings of the 2020 Indo–Taiwan 2nd International Conference on Computing, Analytics and Networks (Indo-Taiwan ICAN), Rajpura, India, 7–15 February 2020, pp. 176–180 (2020)
6. Tang, X., Ghassemlooy, Z., Rajbhandari, S., Popoola, W.O., Lee, C.G.: Coherent heterodyne multilevel polarization shift keying with spatial diversity in a free-space optical turbulence channel. J. Light. Technol. **30**, 2689–2695 (2012)
7. Armghan, A., Alsharari, M., Aliqab, K., Singh, M., Abd El-Mottaleb, S.A.: Performance analysis of hybrid PDM-SACOCDMA-enabled FSO transmission using ZCC codes. Appl. Sci. **13**, 2860 (2023)
8. Ghassemlooy, Z., Popoola, W., Rajbhandari, S.: Optical Wireless Communication: System and Channel Modelling with Matlab. CRC Press, Boca Raton (2019)
9. Singh, M., Kˇríž, J., Kamruzzaman, M.M., Dhasarathan, V., Sharma, A., El-Mottaleb, A., Somia, A.: Design of a high-speed-OFDMSAC-OCDMA-based FSO system using EDW codes for supporting 5G data services and smart city applications. Front. Phys. **10**, 934848 (2022)
10. Sadiku, M.N.O., Musa, S.M., Sudarshan, R.: Free space optical communications: an overview. Eur. Sci. J. **12**, 55 (2016). https://doi.org/10.19044/esj.2016.v12n9p55
11. Bouchet, O., Favennec, P.N.: Wireless Optical Communications. Wiley Online Libr, Hoboken (2012)
12. RAPPORT UIT-R F.2106: Applications du service fixe utilisant des liaisons optiquesn en espace libre (2007)

Enhancing Gain and Bandwidth in 5G Wireless Communication Systems

Ahmed Mohamed Salem[1], Hanane Djellab[2(✉)], Mohamed Lashab[3], and Oumaima Allaoua[1]

[1] Laboratory of LAMIS, Department of Electrical Engineering, University Larbi Tebessi, Tebessa, Tebessa, Algeria
{ahmed.mohamedsalem,oumaima.allaoua}@univ-tebessa.dz

[2] Laboratory of LTI, Department of Electrical Engineering, University Larbi Tebessi, Tebessa, Algeria
hanane.djellab@univ-tebessa.dz

[3] LGEA Laboratory, Department Electrical Engineering, Oum El Bouaghi University, Oum El Bouaghi, Algeria
lashab.mohamed@univ-oeb.dz

Abstract. With the increasing demand for fast and dependable wireless communication, 5G technology has emerged as a highly promising solution. This study addresses the critical objective of enhancing gain and bandwidth performance in 5G wireless communication systems, with particular emphasis on utilizing Multiple-Input Multiple-Output (MIMO) techniques. To achieve our objective, we propose an optimized 2×1 microstrip antenna array design tailored for 5G applications operating in the FR1 frequency range. The antenna array consists of rectangular patch microstrip antennas, with each element measuring 9.6 mm × 13 mm. For accurate simulation and analysis, we employ the HFSS software at a frequency of 6.8 GHz. The selected substrate material is F R4, known for its favorable characteristics, including a relative permittivity ($\epsilon ef f$) of 4.4. Our main goal is to leverage the advantages provided by the proposed design of the antenna array to enhance both gain and bandwidth. By doing so, we aim to optimize the performance of 5G wireless communication systems. The antenna array showcases a significant gain improvement of 6 dB, resulting in expanded bandwidth at 540 MHz. These enhancements are essential for achieving superior coverage, increased data rates, and improved overall system efficiency in the demanding 5G environment.

Keywords: Array patch antenna · gain · bandwidth

1 Introduction

Antennas are essential to the communications industry. They are, to put it simply, machines that transform radio signals into electrical energy and the other way around [1]. The fifth generation (5G) of cellular communication technology is distinguished by its capacity to deliver high data rates (throughput), low latency, support for numerous users

M. R. Laouar et al. (Eds.): ICISAT 2023, LNNS 981, pp. 11–20, 2024.
https://doi.org/10.1007/978-3-031-60591-8_2

in a specific region, effective mobility, and reduced power consumption [2]. The use of antennas with multiple-input multiple-output (MIMO) capabilities, a large number of elements, beamforming capabilities, and wide bandwidth is one of the key technologies that enables the high-speed 5G network [3]. Microstrip antennas are the most popular choice when creating MIMO antennas for 5G applications because to its low profile, simplicity of installation, and small size. Microstrip antennas do, however, have several drawbacks, such as low strength and a limited band-width [4].

Developing MIMO antennas with a large number of elements brings about a notable challenge known as the high mutual coupling effect between adjacent elements. Overcoming this challenge is crucial to achieve desired characteristics such as wide bandwidth, enabling high data rates, and facilitating the use of MIMO antennas in various applications, including Internet of Things (IoT) devices. By effectively addressing the mutual coupling issue through careful design techniques and innovative signal processing approaches, MIMO antennas can harness their full potential, providing enhanced wireless communication capabilities for a range of IoT applications [5].

The remaining sections of this paper are organized as follows. In Sect. 2, we delve into Antenna Design, exploring the considerations, techniques, and tools involved in optimizing the performance of antennas, both for simple antennas and MIMO systems. Section 3 focuses on Antenna Result, where we analyze and compare the performance of simple antennas and MIMO antennas in terms of parameters such as gain, bandwidth, and overall system performance [6]. In Sect. 4, we present a Comparison between the two antenna types, highlighting the advantages and limitations of each approach. Finally, in Sect. 5, we draw a Conclusion summarizing the key findings, discussing the implications of the comparison, and suggesting potential areas for further research in the field of antenna design for both simple and MIMO systems [7].

2 Antenna Design

In this study, a rectangular patch microstrip antenna was designed as a 1×2 array operating at a frequency of 7 GHz. The antenna design utilized two different materials, Cooper and FR-4. The Cooper material was employed for both the patch and ground, while the substrate material utilized was FR-4 with a thickness of 1.575 mm [8].

2.1 Dimensions of Rectangular Patch Microstrip Antenna

The dimensions of the proposed antenna were calculated using well-known formulas for microstrip patch antennas, as outlined below [9].

- The width of the Patch

$$W = \frac{C}{2fr}\sqrt{\frac{2}{\varepsilon_r + 1}} \tag{1}$$

- The length of the patch:

$$L = L_{eff} - 2\Delta L \tag{2}$$

$$Leff = \frac{1}{2fr\sqrt{\mu_0\varepsilon_0}\sqrt{\varepsilon_{eff}}} \tag{3}$$

- The effective refractive index can be expressed as:

$$\varepsilon eff = \frac{\varepsilon r + 1}{2} + \frac{\varepsilon r - 1}{2\sqrt{1 + 12\frac{h}{w}}} \tag{4}$$

- The extension of length of each side can be expressed as:

$$\Delta l = 0.412\frac{(\varepsilon r + 0.3)(\frac{w}{h} + 0.264)}{(\varepsilon r - 0.258)(\frac{w}{h} + 0.8)} \tag{5}$$

The equations mentioned above highlight the significance of various parameters in the design of rectangular microstrip antennas. These parameters play a crucial role in determining the antenna's performance and characteristics [10].

- **Patch** (Fig. 1): $W_p = 13$ mm, $L_p = 9.6$ mm

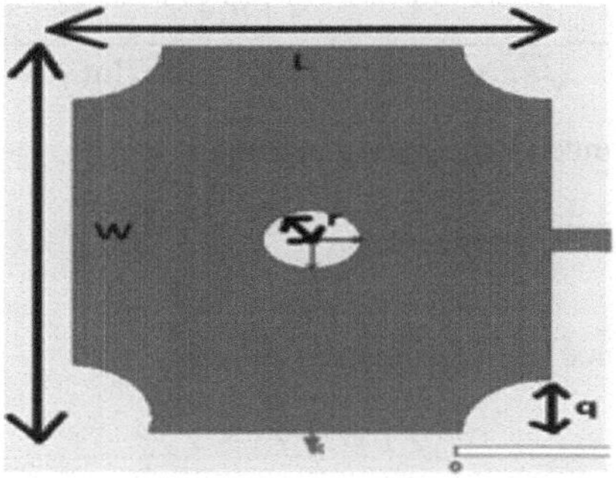

Fig. 1. Patch Dimensions

- **Slots**: $r = 1.4$ mm, $q = 1$ mm
- **Ground** (Fig. 2): $W_s = 41.5$ mm, $L_s = 29$ mm

The ground plane dimension, i.e., length (Lg) and width (Wg) for this design would be given as:

$$\text{Lg} = 6\text{h} + \text{L}$$

$$\text{Wg} = 6\text{h} + \text{W}$$

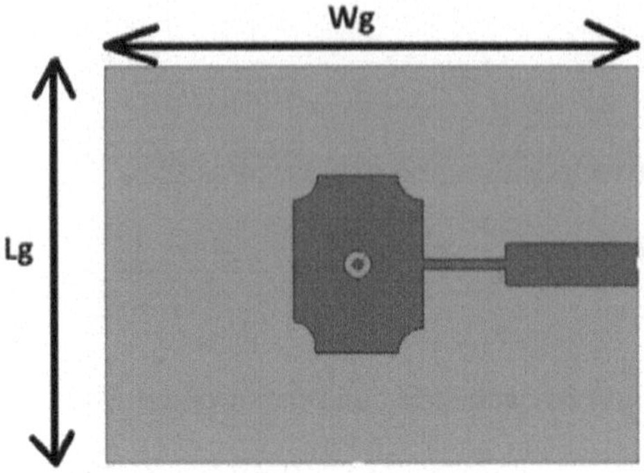

Fig. 2. The Ground

2.2 Feed Dimensions

– Width of transmission line [11]:

$$\text{if } \frac{w}{h} < 1 : z = \frac{60}{\sqrt{\varepsilon r}} \ln\left(\frac{8h}{w} + \frac{w}{4h}\right) > \frac{126}{\sqrt{\varepsilon r}} \tag{6}$$

$$\text{if } \frac{w}{h} > 1 : z = \frac{120\pi}{\sqrt{\varepsilon r} \times \left(\frac{w}{h} + 1.393 + 0.667 \ln\left(\frac{w}{h} + 1.44\right)\right)} < \frac{126}{\sqrt{\varepsilon r}} \tag{7}$$

– The total input impedance of the patch antenna

$$za = 90 \frac{\varepsilon_r^2}{\varepsilon_r - 1} \left(\frac{L}{w}\right)^2 \tag{8}$$

– Characteristic impedance of the transmission line

$$Z1 = \sqrt{Z_2 \times Z_a} \tag{9}$$

whith Z_2: probe impedance

Length of the transmission line is

$$TL = {}^{\lambda_d}\!/_{4} = {}^{\lambda_0}\!\Big/_{4\sqrt{\varepsilon reff}} \tag{10}$$

Feed Dimensions of Patch Antenna
(See Fig. 3 and Table 1)

$$Za = 279.46\Omega$$

Feed Dimensions of 2 × 1 Array Antenna
(See Fig. 4 and Table 2)

$$Za = 197\Omega \; Z1 = 118.2\Omega \; Z2 = 50\Omega \; Z3 = 70.7\Omega \; Z4 = 100\Omega$$

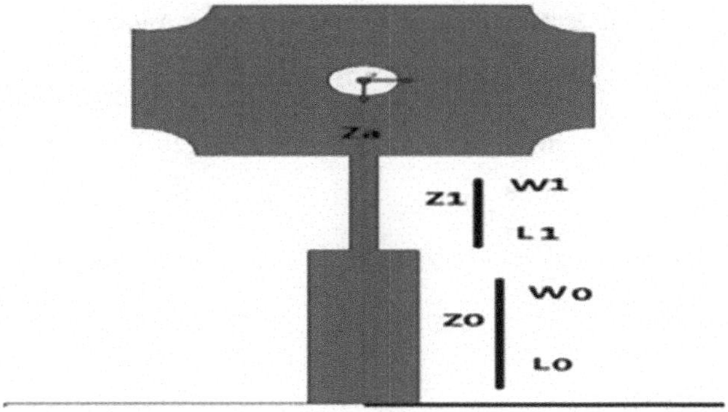

Fig. 3. The patch antenna

Table 1. Feed Dimensions of Patch Antenna

	Width (mm)	Length (mm)
Z1 = 118.2	0.20	2
Z0 = 50	1.56	5.5

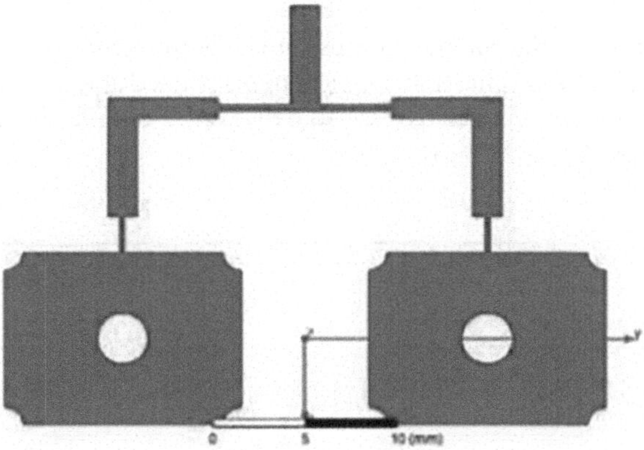

Fig. 4. Array Antenna

Table 2. Feed dimensions of 2 × 1 Array Antenna

	Width (mm)	Length (mm)
Z1 = 118.2	0.2	2
Z2 = 50	1.56	5.5
Z3 = 70.7	1.07	6
Z4 = 100	0.38	4

3 Antenna Result

In contemporary times, it has become a common practice to assess the performance of a system through simulations before its physical realization. This approach finds practical applications, and for our study, we employed the simulator"Ansoft HFSS" to evaluate key parameters such as gain, bandwidth, return loss, and voltage standing wave ratio (VSWR) of the proposed antenna. By leveraging simulation tools like Ansoft HFSS, we were able to reduce fabrication costs by optimizing the design and ensuring its efficacy prior to the manufacturing stage [12].

3.1 Return Loss

To ensure the usability of the antenna, it is essential for the return loss value to be below −10 dB. A lower return loss value indicates a better-designed antenna. In the case of the 2 × 1 array configuration, the return loss is −30.6 dB (Fig. 5), while for the single-element antenna, the return loss is −27.3 dB [13].

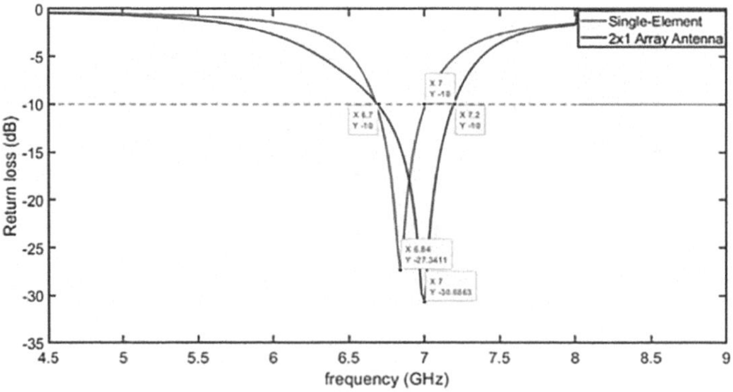

Fig. 5. S11 for patch antenna et 2 × 1 array antenna

3.2 Gain

The gain of an antenna in a specific direction is defined as the ratio of the intensity in that direction to the radiation intensity that would be achieved if the power accepted by the antenna were radiated equally in all directions (isotropically) (Fig. 6) [14].

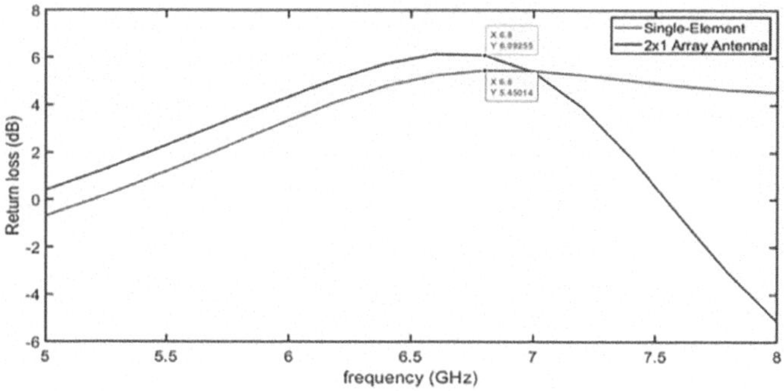

Fig. 6. The gain for the single-element antenna and the 2*1 Array Antenna

From the figure provided, it can be observed that the gain of the array element is 6.86, which is higher than the gain of the single element antenna, which is 4.6. This visual representation reinforces the comparison, illustrating the higher gain achieved by the array element [15].

3.3 Voltage Standing Wave Ratio VSWR

The Fig. 7 illustrates the VSWR response of the antenna. In practical terms, the ideal value for VSWR is less than 2, as it indicates the best performance. The plot in the figure confirms the VSWR values obtained for the antenna and highlights its adherence to the desired performance criteria (Fig. 7) [16].

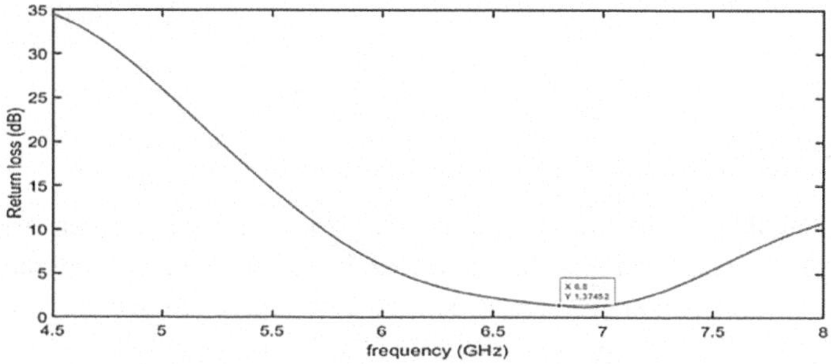

Fig. 7. VSWR for the 2*1 Array Antenna

3.4 Radiation Pattern

(See Fig. 8)

– Single-Element

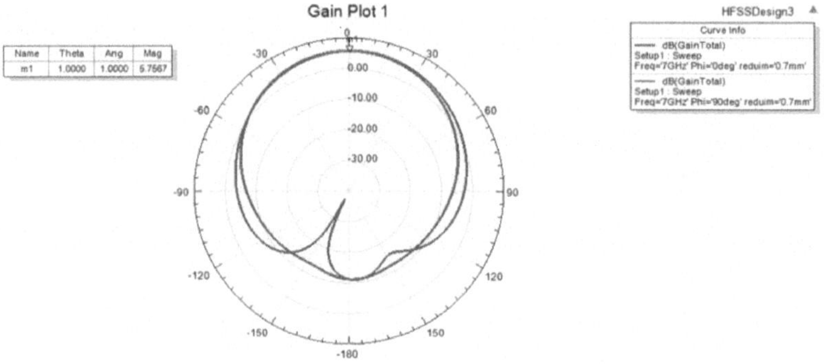

– 2*1 Array Antenna

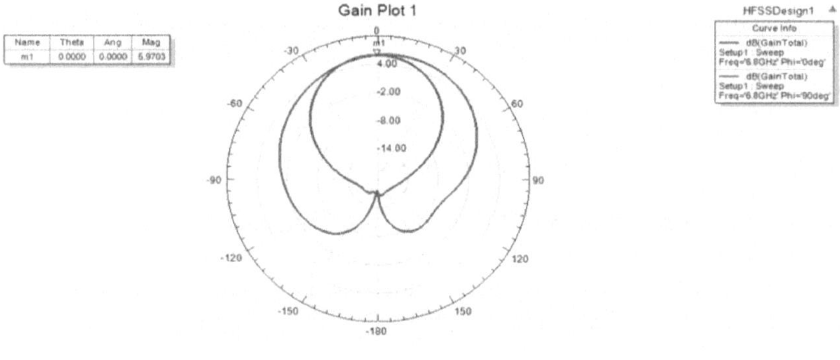

Fig. 8. Radiation Pattern for Array antenna

4 Comparison

When comparing the gain of a single antenna with that of a single element array antenna, we observe that the single antenna has a gain of 5.4, while the array antenna achieves a higher gain of 6. This comparison clearly demonstrates the advantage of utilizing an array configuration, as it results in an increased gain [17]. Moreover, in the context of Internet of Things (IoT) applications, the wider bandwidth of the array antenna (500) becomes even more significant. The expanded bandwidth allows the array antenna to operate over a broader range of frequencies, catering to the diverse communication requirements of IoT devices. This wider bandwidth is particularly valuable in IoT deployments, as it facilitates seamless connectivity and efficient data transfer among a multitude of devices operating on different frequency bands. Therefore, the array antenna's ability to support a wider bandwidth enhances its suitability for IoT applications, enabling versatile and flexible communication capabilities within IoT ecosystems [18] (Table 3).

Table 3. The Gain and Bandwidth of Single Antenna and Array Antenna

	Single Element	Array antenna
Gain (dB)	5.4	6
BandWidth (MHz)	340	560

5 Conclusion

This paper presents the design and analysis results of a rectangular 1 × 2 array and a single patch microstrip antenna operating at 7 GHz, with a focus on their relevance to Internet of Things (IoT) applications [19].

The designed microstrip antenna with a wider bandwidth of 560 MHz offers increased frequency capacity, making it suitable for accommodating various IoT devices operating on different frequency bands. With a VSWR of 1.37, signal reflection was minimized, ensuring optimal transmission. The antenna also demonstrated a gain of 6.0 dB, facilitating effective signal transmission and reception. Comparatively, the single patch microstrip antenna achieved a superior impedance match, It exhibited excellent power transfer characteristics with a VSWR of 1.11. Although it had a narrower bandwidth of 300 MHz, it still provides sufficient capacity for many IoT applications. The antenna achieved a gain of 4.62 dB, show casing its effectiveness in radiating and receiving signals.

These results underscore the significance of considering bandwidth in antenna design for IoT applications. A wider bandwidth enables versatile and flexible communication capabilities, facilitating seamless connectivity and efficient data transfer among IoT devices operating on different frequency bands. The findings contribute to the ongoing optimization and development of antennas for IoT systems, ensuring reliable and efficient wireless communication within IoT ecosystems.

References

1. Song, W.: MIMO antenna array in 5G communication. Highlights Sci. Eng. Technol. **27**, 200135 (2022)
2. Sun, X., Wang, F.: Analysis of MIMO technique in 5G mobile communication array antenna modeling. Changjiang Inf. Commun. 035–004 (2022)
3. Wei, C.: Research on massive MIMO technology in 5G mobile communication. Sci. Technol Wind **16**, 1 (2020)

4. Rana, M.S., Smieee, M.M.R.: Design and analysis of microstrip patch antenna for 5G wireless communication systems. Bull. Electr. Eng. Inform. **11**(6), 3329 (2022)
5. R¨ais¨anen, A.V., Zheng, J., Ala-Laurinaho, J., Karki, S., Viikari, V.: Millimeter-wave antennas for 5G. In: 2018 11th Global Symposium on Millimeter Waves (GSMM)
6. Kishore, S., Rajak, A.: Microstrip patch antenna with C slot for 5G communication at 30 GHz. Emerg. Sci. J. **6**, 1315 (2022)
7. Irfansyah, A., Harianto, B.B., Pambudiyatno, N.: Design of rectangular microstrip antenna 1×2 array for 5G communication. J. Phys. Conf. Ser. **2117**, 012028 (2021)
8. Dhanyswari, A.A., Alam, S., Surjati, I.: Design of dual-band microstrip linear array MIMO antenna with U slot For 5G communication system. J ECOTIPE (Electron. Control Telecommun. Inf. Power Eng.) **10**(1), 32–41 (2023)
9. Rana, M.S., Hossain, S., Rana, S.B., Rahman, M.M.: Microstrip patch antennas for various applications: a review. Indones. J. Electr. Eng. Comput. Sci. **29**(3), 1511–1519 (2023)
10. Sandi, E., Diamah, A., Al Mawaddah, M.: High isolation MIMO antenna for 5G C-band application by using combination of dielectric resonator, electromagnetic bandgap, and defected ground structure. EURASIP J. Wirel. Commun. Network **2022**, 125 (2022). https://doi.org/10.1186/s13638-022-02208-1
11. Bengtsson, E.L., Rusek, F., Malkowsky, S., Tufvesson, F., Karlsson, P.C., Edfors, O.: A simulation frame-work for multiple antenna terminal in 5G massive MIMO systems. IEEE Access **5**, 26819–26831 (2017)
12. Swindlehurst, A.L., Ayanoglu, E., Heydari, P., Capolino, F.: Millimeter-wave massive MIMO: the next generation revolution. IEEE Commun. Mag. **52**, 56–62 (2014)
13. Garg, R.: Microstrip Antenna Design Handbook. Artech House, Boston (2000)
14. Gao, Y., Ma, R., Wang, Y., Zhang, Q., Parini, C.: Stacked patch antenna with dual-polarization and low mutual coupling for massive MIMO. IEEE Trans. Antennas Propag. **64**(1), 4544–4549 (2016)
15. Khraisat, Y.S.H.: Design of 4 elements rectangular microstrip patch antenna with high gain for 2.4 GHz applications. Mod. Appl. Sci. **6**(1), 68 (2012). https://doi.org/10.5539/mas.v6n1p68
16. Dahlman, E., et al.: 5G wireless access: requirements and realization. IEEE Commun. Mag. **52**, 42–47 (2014)
17. Bajpai, R.K., Paulus, R., Singh, A., Aneesh, M.: Review: dual band microstrip antennas for wireless applications. Int. J. Adv. Appl. Sci. (IJAAS) **7**(2), 143–151 (2018). https://doi.org/10.11591/ijaas.v7.i2.pp143-151
18. Constantine, A., Balanis, A.: Antenna Theory Analysis and Design, 4th edn. Wiley, Hoboken (1938)
19. Prasanna Venkatesan, G.K.D.: Optımısatıon of MIMO Antenna for 5G Applıcatıons. Karpagam Academy of Higher Education, Coimbatore (2018)

Boosting Regression Assistive Predictive Maintenance of the Aircraft Engine with Random-Sampling Based Class Balancing

Ibrahima Barry[1], Meriem Hafsi[1](✉), and Saeed Mian Qaisar[1,2]

[1] CESI LINEACT, 15C Av. Albert Einstein, 69100 Villeurbanne, France
{ibarry,mhafsi,smianqaisar}@cesi.fr

[2] Electrical and Computer Engineering Department, Effat University, Jeddah 22332, Kingdom of Saudi Arabia

Abstract. This study presents the development of a data-driven predictive maintenance model in the context of industry 4.0. The solution is based on a novel hybridization of Remaining Useful Life (RUL) generation, Min-Max normalization, random-sampling based class balancing, and XGBoost regressor. The applicability is tested using the NASA's C-MAPSS dataset, which contains aircraft engine simulation data. The objective is to develop an effective and Artificial Intelligence (AI) assistive automated aircraft engine's RUL predictor. It can maximize the benefits of predictive maintenance. The rules based RUL generation provides a ground truth for evaluating the performance of intended regressors. The Min-Max normalization linearly transforms the intended dataset and scales the multi subject's data in a common range. The imbalance presentation among intended classes can lead towards a biasness in findings. This issue is intelligently resolved using the uniformly distributed random sub-sampling. Onward, the performance of robust machine learning and ensemble learning algorithms is compared for predicting the RUL of the considered aircraft engine by processing the balanced dataset. The results have shown that the XGBoost regressor, uses an ensemble of decision trees, outperforms other considered models. The root mean square error (RMSE) and mean absolute error (MAE) indicators will be used to evaluate the prediction performances. The devised method secures the RMSE value of 12.88%. It confirms a similar or better performance compared to the state-of-the-art counterparts.

Keywords: XGBoost · Random-Sampling · Class Balancing · Remaining Useful Life (RUL) · C-MAPSS · Machine learning · Ensemble learning

Abbreviations

AI	Artificial Intelligence
CBM	Condition-Based Maintenance
CNN	Convolutional Neural Networks
DL	Deep Learning
DM	Data Mining

M. R. Laouar et al. (Eds.): ICISAT 2023, LNNS 981, pp. 21–36, 2024.
https://doi.org/10.1007/978-3-031-60591-8_3

IoT	Internet of Things
KNN	K-Nearest Neighbor
XGBoost	EXtreme Gradient Boosting
LSTM	Long Short-Term Memory
ML	Machine Learning
PdM	Predictive Maintenance
PHM	Prognostics and Health Management
RUL	Remaining Useful Life

1 Introduction

Over the past few decades, the manufacturing and industrial sector has witnessed continuous evolution and growth. We have just entered the era of Industry 4.0, and yet we are already beginning to explore and envision the prospects of Industry 5.0.

In this constant evolution [1], companies must proactively adapt to changes to ensure their competitiveness, particularly in the industrial maintenance, which plays a crucial role. Indeed, for optimal levels of quality and profitability in production, it is imperative to maintain the equipment in an optimal operating condition [2]. Therefore, the main objective of industrial maintenance is to ensure the availability, reliability and efficiency of machines and components, minimizing unplanned downtime and optimizing production performance [3]. In this context, ensuring and maintaining the health status of equipments and their components is a significant challenge in complex industrial systems where fulfilling their function relies on ensuring their reliability and efficiency [4]. Prognostics and Health Management (PHM) is a discipline that focuses on understanding the degradation mechanisms of systems, aiming to estimate their health status and proactively anticipate potential failures [5]. Its ultimate objective is to optimize maintenance tasks, ensuring improved system performance and reliability. PHM encompasses different methods including monitoring, detection of anomalies, diagnostics of causes, prognosis of remaining useful life (RUL) and optimization of maintenance [6]. It thus makes it possible to monitor the health state in real time and to provide relevant information for decision support on the actions to be implemented to maintain the system in optimal operational conditions. The integration of the Internet of Things (IoT) and hyper-connectivity of equipment in the industry has catapulted it into a new era [3], where the massive data generated represents a formidable treasure and offers a wide array of possibilities. At the same time, the integration of advanced technologies such as automation, exponential computing capabilities, data analysis and machine learning, industrial maintenance aims to prevent breakdowns and failures [4], anticipate repair needs and maximize the life of assets, enabling companies to remain competitive and promoting their productivity and efficiency [7]. Predictive Maintenance (PdM)[1, 6] is one of the most promising applications of PHM which aims to predict failures and errors by estimating the RUL of a system or equipment.

Over the past few decades, the literature has put forth various methods and approaches – based on physics, data or experiences – [3] with diverse applications in manufacturing, telemetry [6], automotive, railway, aviation and aeronotic [2], centrifugal compressor industry, industrial rotating machines and semiconductor, batteries life estimation [8] and others domains [1]. Aircraft engine life estimation [2] is one of the applications that have aroused growing interest in recent years, notably accentuated by the accessibility of the C-MAPSS (Commercial Modular Aero-Propulsion System Simulation turbofan engine) dataset provided by NASA[1] for scientific communities working on the RUL prediction and development of new methods [9]. Indeed, aircraft engines are crucial assets with stringent reliability requirements and demanding airworthiness conditions. They face significant challenges as any unforeseen degradation can result in substantial operational disruptions, delays, costly downtime, and potentially even fatal accidents [10]. In the aerospace industry, maintenance is undergoing a swift transformation, moving from preventive strategy to condition-based maintenance (CBM) [5] and predictive maintenance (PdM). In traditional preventive maintenance, maintenance actions such as replacements, inspections, and data collection are scheduled based on fixed time intervals [3]. However, this method lacks the ability to assess each asset individually as their behavior may vary under different contexts or operating conditions [2]. PdM seeks to address this limitation by predicting the state and reliability of each asset, estimating its time to failure [10]. PdM aims to incorporate reliable RUL prognostic into maintenance planning effectively [9]. The literature is replete with data-driven works to estimate the RUL of aircraft engines. However, the machine learningbased approaches proposed lack rigor, and the results are not thoroughly analyzed, leading to a lack of continuation in the research. Additionally, while few works employ advanced conventional machine learning methods such as KNN (K Neareast Neighbors), RF (Random Forest), SVM (Support Vector Machine) [4, 6], many authors favor DL algorithms for its superior performance [9]. Nonetheless, the interpretation and explainability of results is often absent [3], likely due to the black box aspect of DL. Consequently, the prevailing observation is that the data-driven approach dominates the current research context [1].

In this article, we address the challenges of predictive maintenance for aircraft engines, focusing specifically on the RUL prognostic using ML techniques. Based on the issues highlighted in the literature, we propose a novel method using conventional ML methods that offer both high performance and improved interpretability of the training model.

1.1 Background

1.1.1 RUL Prediction

According to the previous section, PdM employs advanced data analysis techniques and IA to predict failures before they occur by monitoring early warning signals of malfunction. it involves predicting the RUL of a component of equipment before it experiences a critical failure. This helps in effectively scheduling maintenance, reducing downtime, and preventing unexpected breakdowns [2].

The RUL concept (Fig. 1) is based on the recognition that equipment undergoes progressive wear and tear over time, which can lead to failures or malfunctions, by monitoring equipment operating parameters in real-time, such as vibrations, temperature, pressure, and electrical currents, and using sophisticated algorithms to analyze this data,

[1] C-MAPSS NASA Dataset: https://data.nasa.gov/dataset/C-MAPSS-Aircraft-Engine-Simulator-Data/xaut-bemq.

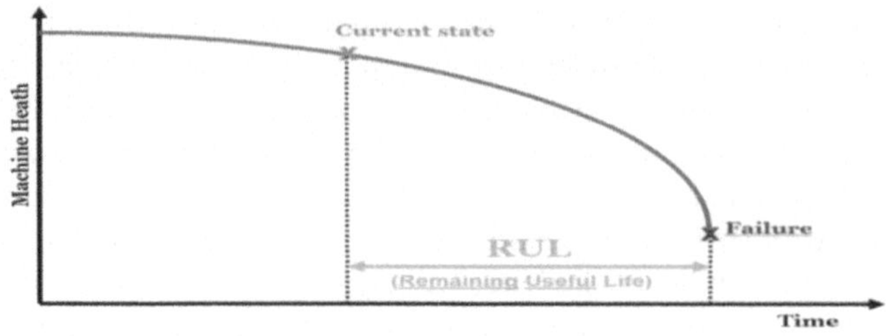

Fig. 1. System degradation over time and RUL estimation which represents the duration a system can continue operating before it needs to undergo repair or replacement due to failure.

collected from multiple sensors, it is possible to predict with some accuracy the point at which a piece of equipment is likely to fail [4].

1.1.2 PdM Approaches

Four approaches are mainly used in the literature for RUL prediction: physical model-based, knowledge-based model, data-driven model and hybrid model [1, 11]:

- **Physical-based approach:** describes the physical degradation process and fault propagation with mathematical models [3]. It is particularly useful for understanding complex physical phenomena and where data-driven models are not sufficient due to the lack of data, but it can be complex, computationally expensive, inflexible, and require data calibration, updates, and significant initial investment [10].
- **Data-driven approach:** aims to detect the degradation evolution and forecast the future health state of a system by leveraging historical data [11], and relies on techniques such as statistical analysis, ML, and DM to extract valuable insights. It is particularly adept at detecting patterns and uncovering hidden correlations, leading to enhanced accuracy in failure predictions. Nevertheless, the lack of data, system complexity, and varying operational conditions limit its widespread application. Finally, the lack of explainability of ML-based models represent a major limitation [12].
- **Knowledge-based approach:** utilizes human expertise and equipment-specific knowledge to anticipate failures, by integrating and formalizing field observations, diagnostic rules, and technician experience [3]. It offer an explicative results and enhances data quality, models causal relationships, adapts to changing operational conditions and reduces dependence on large volumes of data. By incorporating domain expertise, it fills the gaps in data-driven approaches, leading to more accurate models and better failure anticipation [5]. However, it may face some limitations, such as the difficulty of obtaining precise knowledge for predictability pruposes from experience and limited access to expert sources.
- **Hybrid approach:** combines two or more of the previous PdM approach for increasing predictive accuracy [5, 12].

In this present work, we are particularly interested in the data-driven approach using conventional ML applied to aircraft engine data. The next section discusses existing work in this area (Table 1).

Table 1. Comparison of PdM Approaches

Approach	Advantages	Disadvantages
Physics-based	Without historical data, reliable	Complexity of systems
Data-driven	High accuracy	Requires a large amount of data, interpretation
Knowledge-based	Effective for simplified cases, explainability	Low accuracy, limited in complex systems
Hybrid	High Accuracy, explainability	Hard to buid, fusion and incertainty management

1.2 Related Works

Several works on RUL estimation of aircraft engines have been carried out since the release of the C-MAPSS dataset [2]. Many of these works have utilized DL algorithms [13–24], considering them the most appropriate for this dataset: Sateesh Babu et al. [13] employed Convolutional Neural Networks (CNN) to RUL estimation and de Pater et al. [9] developed an alarm-based maintenance planning framework that integrates RUL prognostic based on CNN in the maintenance schedule. Zheng et al. [14] implemented Long Short-Term Memory (LSTM) networks in their research and Wang et al. [15] opted for Bidirectional LSTM (BiLSTM) in their analysis. Liu et al. [22] applied an Attention-based Bidirectional LSTM (Bi-LSTM) method. Some works have combined several approaches like CNN and LSTM [17] or BiLSTM [25]. Mo et al. [18] developed a multi-head architecture combines CNN and LSTM, Kumar [20] introduced a hybrid model for their analysis, Asif et al. [23] merged LSTM with Automatic Piecewise Linear Modeling, and Peng et al. [26] combined LSTM with a Fully Connected Layer CNN (FCLCNN) in their research. Other authors have proposed original methods such as Adaptive Gradient CNN (AGCNN) in [19], a Broad Learning System (BLS) coupled with a Temporal Convolutional Network (TCN) is presented in [21], Li et al. [16] integrated Directed Acyclic Graphs (DAG) in their methodology and Liu et al. [22] applied an Attentionbased Bidirectional LSTM (Bi-LSTM) method. Finally, Hu et al. [24] applied a Deep Reinforcement Learning approach for RUL estimation (DRLRULe) in their investigation. However, in this particular study, we demonstrate that traditional ML algorithms can also be highly effective.

1.3 Contribution

In this paper, a novel hybridization of RUL generation, Min-Max normalization, random-sampling based class balancing, and XGBoost regressor is proposed for an automated aircraft engine's RUL prediction.

- The rules based RUL generation provides a ground truth for evaluating the performance of intended regressors.
- The Min-Max normalization linearly transforms the intended dataset and scales the multi subject's data in a common range.
- The imbalance presentation among intended classes can lead towards a biasness in findings. This issue is intelligently resolved using the uniformly distributed random sub-sampling.
- The performance of robust ML and ensemble learning algorithms is compared for predicting the RUL of the considered aircraft engine by processing the balanced dataset.

The results have shown that the XGBoost regressor outperforms other considered models.
- The devised method secures the RMSE value of 12.88%. It confirms a similar or better performance compared to the state-of-the-art counterparts.

The rest of the paper is organized as follow: Sect. 2 describes the used materials and methods. The findings are presented and described in Sect. 3. Finally, the conclusion is made in Sect. 4.

2 Materials and Methods

2.1 Methodology

To answer the problem mentioned above, we have implemented our solution according to the functional diagram (Fig. 2) and providing an overview of key system components and interconnections, demonstrating how data flows through the system from one stage to another, and how the components interact with each other. All these steps will be detailed in the following sections.

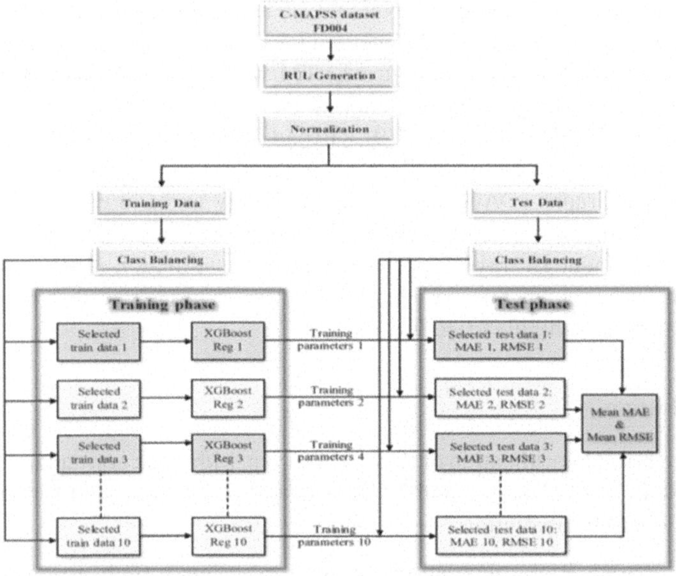

Fig. 2. Proposed framework RUL prediction based on Random Sampling Based Class Balancing Method.

2.2 Dataset

C-MAPSS (Commercial Modular Aeropropulsion System Simulation) is provided by NASA (National Aeronautics and Space Administration) and was first made available in 2008 during the International Conference on PHM (Prognostics and Health Management). Since its release, it has been extensively researched to enhance the accuracy of RUL prediction, with the ultimate goal of planning component maintenance before failure occurs [27]. C-MAPSS dataset is widely recognized as the benchmark dataset for RUL prognostic. Moreover, its characteristics enable the findings to be applicable to various other research and industrial use cases. The dataset includes multivariate time series augmented with metadata on operational settings. Additionally, it encompasses multiple physical interactions among the recorded properties. The dataset also covers various fault modes and operating conditions, while considering manufacturing variations, distinct lifespans, and natural failures [2]. C-MAPSS is subdivised into four units: FD001, FD002, FD003, and FD004 with different multivariate time series and various numbers of engines. Each unit contains three files: training data, testing data and RUL data. Table 2 describes caracteristics of each unit. The dataset is composed of 21 sensors with the application of three different operating conditions in relation to the cycle time for each engine. A turbojet engine is mainly composed of: Low Pressure Turbine (LPT), High Pressure Turbine (HPT), Fan, Nozzle, Low Pressure Chamber (LPC), High Pressure Chamber (HPC), etc. (Fig. 3) shows how these different components and the way they are interconnected.

Table 2. Description of the properties and different batches of the C-MAPPS dataset.

	FD001 Unit	FD002 Unit	FD003 Unit	FD004 Unit
N° Engines available in train set	100	260	100	249
N° Engines available in test set	100	259	100	248
Operating conditions applied	1	6	1	6
Fault modes	1	1	2	2
Training set size	20,632	53,760	24,721	61,250
Testing set size	13,097	33,992	16,597	41,215

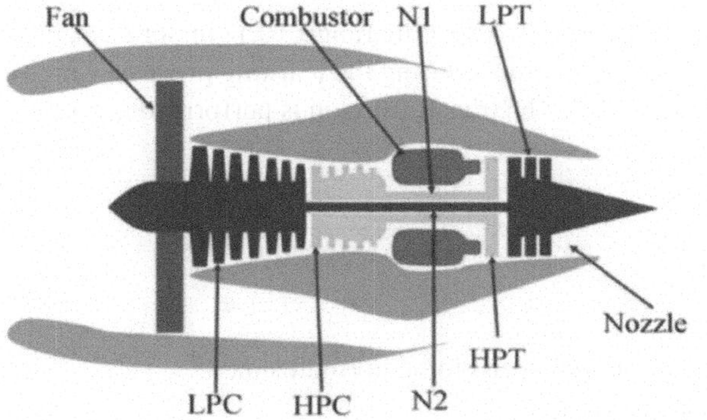

Fig. 3. Simplified C-MAPSS turbofan engine diagram [28].

According to the data description in Fig. 2, units FD002 and FD004 exceptionally contain 6 operating conditions with more than 200 tracking trajectories. Asif et al. demonstrated in [23] that the two subunits possess irregular behavior, making the prediction of RUL for these complex sequences a challenging task. As a result, most works in the literature concentrate on the FD001 and FD003 datasets due to their simplicity and smaller data volume. Consequently, the FD004 dataset remains relatively underexplored in existing literature.

This paper presents preliminary findings for the subset FD004. In future a detailed study will be presented, in an upcoming journal paper, while considering all subsets of the C-MAPSS dataset.

2.3 Remaining Useful Life (RUL) Generation

Each sub-dataset is composed of 26 columns in total, but the target RUL column is not included. The RUL is determined by calculating the difference between the maximum time (in cycle) for a given unit and the current time (in cycle). To achieve this, we proceeded as follows:

$$RUL_{train}(i) = max(Cycle)_{per\ unit}(i) - Cycle(i) \tag{1}$$

$$RUL_{test}(i) = max(Cycle)_{per\ unit}(i) + RUL_{given}(i) - Cycle(i) \tag{2}$$

Where:

- $RUL_{train}(i)$ and $RUL_{test}(i)$ are the *RUL* for the train and test sets respectively,
- $max(Cycle)_{per\ unit}(i)$ is the maximum number of cycles for each unit i,
- $Cycle(i)$ is the current cycle number,
- $RUL_{given}(i)$ is the given cycles for each unit i in the RUL file.

2.4 Min-Max Normalization

Min-Max Normalization, also known as feature scaling, is a simple method used in data preprocessing to scale numerical data. The goal of Min-Max Normalization is to transform features to fall within a specific range, [0,1] in our case, thus ensuring that no particular feature dominates others during the learning process, which can be crucial for some MLg algorithms [23]. The transformation is performed according to the following formula:

$$\mathrm{x}' = \frac{\mathrm{x} - \min(\mathrm{x})}{\max(\mathrm{x}) - \min(\mathrm{x})} \tag{3}$$

Where:

- x is the current value that needs to be normalized.,
- x' is the normalized value,
- $min(x)$ and $max(x)$ correspond to the minimum and the maximum values in the dataset, respectively.

2.5 Random-Sampling Based Class Balancing

The performance of ML algorithms can be affected because of the ample imbalances in the intended classes [29]. This problem can be resolved to a certain extent using the subsampling, oversampling or reweighting approaches [30]. In literature, it is shown that the class balancing can raise the performance of contemporary classification algorithms in terms of accuracy and latency [31, 32].

In this study, we have devised a random sub-sampling approach for class balancing. It resolves the problem of data misbalancing while diminishing the impact of biasing. Moreover, it also diminishes the computational load and latency of the considered classifiers during the training and testing phases and allows to compute empirical results using affordable personal computers.

The approach is based on the uniformly distributed random indexes generation. Let $C = \{C_1, C_2, \ldots, C_K\}$ is a set of intended classes, where K is presenting the considered count of classes. Let the count of instances for these classes is defined by the set $M = \{M_1, M_2, \ldots, M_K\}$ and their corresponding indexes ranges are defined as: $R = \{R_1, R_2, \ldots, R_K\}$. Where, elements of sets M and R belongs to the set of positive integers. Then using the inversion method, based on the idea that "Continuous Cumulative Distribution Functions" (CDFS) have a uniform distribution over the open interval [0,1], the set of uniformly distributed random indexes $I_1 = \{n_1, n_2, \ldots, n_N\}$ is generated for C_1 [33]. A similar process is followed for the generation of $I_2, I_3, \ldots, I_K$ which are sets of the random indexes respectively for the $C_2, C_3, \ldots, C_K$.

The generation process of I_1 is achieved while respecting the corresponding index range, R_1. In this scenario, If x is a uniform random number on $[0,1]$, then a random index i_m, from the standard normal distribution and the specified CDF, F, is generated as: $i_m = F^{-1}(x)$ [33]. Where m ranges from $[1,N]$. In next step, the value of the generated im is adapted as per R_1, the process is given as: $n_m = OS + round(i_m \times SF)$, where SF is the scaling factor and OS is the off-set. The values of SF and OS can be specific as a function of the intended range from the set R. Also, for an effective class balancing, the value of N should be less than or equal to the minimum element in the set M. In this study, the value of $N_{Training} = 50$ is chosen and the value of $N_{Testing} = 25$ is selected.

Algorithm 1: Random-Sampling Based Class Balancing Algorithm - RSBCB

Input : Set of intended classes $C = \{C_1, C_2, ..., C_K\}$;
Count of instances for each class $M = \{M_1, M_2, ..., M_K\}$;
Index ranges for each class $R = \{R_1, R_2, ..., R_K\}$;
Scaling factor SF;
Offset OS;
Number of training samples N_{Training};
Number of testing samples N_{Testing};
Output: Randomly sampled dataset for class balancing

for *each class* C_i *in* C **do**
  Generate uniformly distributed random indexes $I_i = \{n_1, n_2, ..., n_N\}$ for C_i within the index range R_i;
  for *each index* n_m *in* I_i **do**
    Generate a random number x as a uniform random number on $[0, 1]$;
    Generate a random index i_m from the standard normal distribution and the specified CDF F using the inverse method: $i_m = F^{-1}(x)$;
    Adapt the value of i_m as per the index range R_i: $n_m = OS + \text{round}(i_m \times SF)$;
  end
  Select a random sample of N_{Training} instances from class C_i using the generated indexes I_i;
  Store the selected instances in the training dataset;
end
for *each class* C_i *in* C **do**
  Select a random sample of N_{Testing} instances from each class C_i using the generated indexes I_i;
  Store the selected instances in the testing dataset;
end

2.6 Boosting Based Regression

Designed to have specific properties such as efficiency, flexibility and portability, XGBoost (eXtreme Gradient Boosting) Regressor is an optimized distributed gradient boosting library. By implenting ML algorthms under the Gradient Boosting framework, it provides a parallel tree boosting (commonly known GBDT or GBM) allowing to solve many data science problems quickly and precisely [34]. The core XGBoost model is an ensemble of decision trees, where new trees are added to correct the errors made by existing trees. The trees are added sequentially until no further improvements can be made.

Given a set of n instances $(x_1, y_1),\ldots,(x_n, y_n)$ where $x_i \in R^m$ is a feature vector and $y_i \in R$ is the corresponding label, XGBoost learns a function f: $R^m \rightarrow R$ that maps a feature vector to a label. The function f is a series of K decision trees.

$$f(x) = \sum_{k=1}^{K} f_k(x) \tag{4}$$

Where f_k is a decision tree.

The objective of XGBoost is to minimize the following loss function:

$$L(y, f(x)) = \sum_{=i1}^{n} l(y_i, f(x_i)) + \sum_{k=1}^{K} \Omega(f_k) \tag{5}$$

Where l is a differentiable convex loss function that measures the difference between the prediction $f(x_i)$ and the correct label y_i, and Ω is a regularization term that penalizes the model complexity.

XGBoost offers several advantages as a fast and accurate method, but it also includes several regularizations which help to reduce overfitting. Moreover, it supports missing values, allows for parallelizable processing and provides several flexible features for user customization. Finally, several applications are considered: regression solving, ranking, classification, user-defined prediction problems, etc.

Algorithm 2: XGBoost Regressor Algorithm

```
Function my_XGB_Regressor(x, y):
    Input  : Training data x, Testing data y
    Output: None
    mse_test_list ← [],
    rmse_test_list ← [],
    mae_test_list ← []
    for j in range(1, 11) do
        normalized_train_df ← normalised_df(x)
        normalized_test_df ← normalised_df(y)
        train_selected ← Class_Balancing_RSBCB(normalized_train_df, N_Training , j)
        test_selected ← Class_Balancing_RSBCB(normalized_test_df, N_Testing, j)
        X_train, Y_train, X_test, Y_test ← data_split(train_selected, test_selected)
        // Create the XGBoost Regressor model
        model ← XGBRegressor(n_estimators=100, max_depth=3, learning_rate=0.01,
         subsample=0.5, colsample_bytree=0.5)
        model.fit(X_train, Y_train)
        // Test
        y_pred_test ← model.predict(X_test)
        mse_test ← mean_squared_error(Y_test, y_pred_test)
        rmse_test ← sqrt(mse_test)
        mae_test ← mean_absolute_error(Y_test, y_pred_test)
        mse_test_list.append(mse_test)
        rmse_test_list.append(rmse_test)
        mae_test_list.append(mae_test)
        // Print results: mae_test, rmse_test
    end
    mean_mae ← mean(mae_test_list)
    mean_rmse ← mean(rmse_test_list)
    // Print Mean MAE and Mean RMSE
    print("Mean MAE:", mean_mae)
    print("Mean RMSE:", mean_rmse)
```

Following the RUL generation and data normalization, the next stage is to proceed to ten iterations. During each iteration, a random sampling-based class balancing operation is performed. Subsequently, the Mean Absolute Error (MAE) and the Root Mean Squared Error (RMSE) are calculated to evaluate the performance of the model. These two error metrics are then stored and accumulated throughout the 10 iterations. At the end of the 10 iterations, the algorithm computes the average of the 10 MAE and RMSE values. These averages represent the global error rates of our model. In simpler terms, the algorithm performs 10 learning and evaluation experiments, and it utilizes the outcomes to deliver an overall performance evaluation of the model.

2.7 Evaluation Measures (RMSE and MAE)

In the domain of ML, a metric refers to a function employed to quantify the performance of a model or algorithm. In essence, metrics are utilized to measure how accurately a model's predictions align with the actual data [4]. There are several metrics divided into two categories based on the type of problem: classification problems and regression problems. In the context of regression, two of the most commonly used metrics are the MAE (Mean Absolute Error) and the RMSE (Root Mean Squared Error). Both MAE and RMSE express average model prediction error in units of the variable of interestn these two scores are said to be negatively-oriented, meaning that lower values are better.

1. **Mean Absolute Error (MAE):** measures the average magnitude of the errors in a set of predictions, regardless of their direction. It is can be considered as the average over the test sample of the absolute differences between prediction and actual observation where all individual differences have equal weight as shown in Formula (6). MAE is less sensitive to outliers compared to RMSE [35].
2. **Root Mean Squared Error (RMSE):** is a quadratic scoring rule that also measures the average magnitude of the error. It is based on the square root of the average of squared differences between prediction and actual observation. The RMSE gives a relatively high weight to large errors because the squared large difference of prediction and the actual observation leads to even larger contributions to the mean as shown in Formula (7). The effect is to pay more attention to large errors [35].

$$MAE = \frac{1}{n}\sum_{i=1}^{n}|y_i - \hat{y_i}| \tag{6}$$

$$RMSE = \sqrt{\frac{1}{n}\sum_{i=1}^{n}(y_i - \hat{y_i})^2} \tag{7}$$

Where:

- y_i is the actual RUL value of the aircracft engine,
- $\hat{y_i}$ is the predicted RUL value by the model,
- n is the total number of observations.

3 Results and Discussion

In this work, two regression models namely, the KNN and XGBoost are considered for processing the subset FD004 – The KNN Regressor algorithm is well detailed in the article [36] – In our study, the work carried out with the KNN allowed us to find the most optimal parameters described as follows: best number of neighbors 46 and the distance metric as Manhattan distance in the 5th iteration out of a total of 10 tuning iterations with the class balancing. The performance of the considered regressors is evaluated for two different scenarios. In first case, the intended dataset is processed only after normalization. However, in the second case, the intended dataset is normalized and onward the class balancing is performed using the devised random-sampling based approach. The findings are outlined in Table 3.

In the first case the XGBoost regressor attains 19.22% RMSE and 14.48% MAE values respectively. KNN regressor performs better than the XGBoost regressor in this scenario by securing 15.71% RMSE and 11.58% MAE values respectively. In the second case the KNN regressor attains 13.10% RMSE and 09.55% MAE values respectively. XGBoost regressor outperforms by securing 12.88% RMSE value.

However, the MAE value of the KNN regressor is 0.52% lesser than the MAE score achieved with the XGBoost regressor. These results confirm the benefit of using random-sampling based class balancing on the normalized data before conveying it to the regression stage. For both considered regressors the accuracy is improved for the aforementioned second case. In the case of KNN regressor the respective error

Table 3. Evaluation Measures of Considered Regressors: KNN and XGBoost.

Method	FD004	
	RMSE	MAE
KNN Regressor + Normalization	15.71	11.58
KNN Reg + Normalization + Random-Sampling Based Class Balancing	13.10	**9.55**
XGBoost Reg + Normalization	19.22	14.48
XGBoost Reg + Normalization + Random-Sampling Based Class Balancing	**12.88**	10.07

reduction in RMSE and MAE is 02.61% and 02.03%. In the case of XGBoost regressor the respective error reduction in RMSE and MAE is 06.34% and 04.41%.

In order to evaluate and compare the results of our study with litterature works, we used the RMSE metric which is commonly used in the majority of studies conducted on the C-MAPSS dataset [13–24]. The results obtained with the tested methods are displayed in Table 3, they show that XGBoost Regressor model with normalization and Random-Sampling Based Class Balancing achieved the best result based on RMSE. Table 4 presents a comprehensive comparison of innovative and successful methods from the literature including our method. The table includes 13 different methods, their respective preprocessing steps, and a performance comparison based on their RMSE values.

The methods include a range of techniques based on different types of neural networks (CNN, LSTM, BiLSTM), hybrid models, and other advanced ML techniques. Each method is accompanied by specific preprocessing steps such as normalization, feature selection, RUL target function, and others. The best RMSE from prior state-of-the-art methods is 16.96% using Bi-LSTM + Attention [22], while the worst RMSE is 29.15% from CNN [13]. Our proposed method achieves an RMSE of 12.88%, which is 4.08% lower than the best previous method and 16.27% lower than the worst previous method. This underlines the meaningful advancements in RUL prediction accuracy gained by the proposed method over current techniques. The results validate the efficacy of the proposed innovations in improving model performance.

Table 4. Performance comparison with state-of-the-art counterparts and preprocessing steps detailed.

N°	Methods	Author	Year	Normalization	Feature Selection	RUL Target Function	Piece-wise function	Variance threshold	Heath indicator	Correlation analysis	Median move filter	STW Proc.	RUL Generation	RSBCB	RMSE
1	CNN	Sateesh Babu et al.[13]	2016	x		x									29.15
2	LSTM	Zheng et al.[14]	2017	x		x									28.17
3	BiLSTM	Wang et al.[15]	2018	x	x	x									24.86
4	DAG	Li et al.[16]	2019	x	x		x								22.43
5	CNN+LSTM	Kong et al.[17]	2019	x				x	x						23.25
6	Multi-head CNN+LSTM	Mo et al.[18]	2020		x	x									22.89
7	AGCNN	Liu et al.[19]	2020	x	x	x									21.15
8	Hybrid model	Kumar[20]	2021	x	x		x								22.32
9	BLS+TCN	Yu et al.[21]	2022												18.12
10	BiLSTM+Attention meth	Liu et al.[22]	2022			x									16.96
11	LSTM+Auto piecewise	Asif et al.[23]	2022	x		x	x			x	x				17.63
12	DRLRULe	Hu et al.[24]	2023	x	x							x			18.87
13	**Proposed approach**	**Barry et al.**	**2023**	x									x	x	**12.88**

Legend:
STW Proc.: *Sliding Time Window Processing*
RSBCB: *Random Sampling Based Class Balancing*

4 Conclusion

In this paper, a novel data-driven predictive maintenance model is presented in the context of industry 4.0. The proposed method is based on a new fusion of XGBoost regressor, Min-Max normalization, RandomSampling based class balancing, and Remaining Useful Life (RUL) generation. The performance of the intended regressors is assessed using the ground truth, computed by the rules-based RUL generation. The Min-Max normalization adjusts the data from several subjects in a common range by performing the linear transformation. An unbalanced presentation of the intended classes may result in biased findings. The uniformly distributed random sub-sampling is used to intelligently handle this problem. It not only diminishes the biasness in findings but also relaxes the requirements of computational resources.

The performance of considered robust machine learning and ensemble learning algorithms is compared for predicting the RUL of the considered aircraft engine by processing the balanced dataset. The results have shown that the XGBoost regressor outperforms other considered models. The devised method secures the Root Mean Squared Error (RMSE) value of 12.88%. It is shown that the proposed approach secured a similar or better performance compared to the state-of-the-art counterparts.

In this study, the performance of devised approach is evaluated only for the subset FD004. In future, a detailed analysis will be presented while considering all available subsets in the C-MAPSS dataset. Moreover, the incorporation of feature selection and deep learning approaches can enhance the performance of suggested solution. The feasibility of this concept will be investigated in future.

References

1. Hafsi, M., Hamour, N., Ouchani, S.: Predictive maintenance for smart industrial systems: A roadmap. In: Procedia Computer Science. The 6th International Conference on Emerging Data and Industry 4.0 (EDI40), vol. 220, pp. 645–650 (2023)
2. Vollert, S., Theissler, A.: Challenges of machine learning-based rul prognosis: a review on NASA's C-MAPSS data set (2021). https://doi.org/10.1109/ETFA45728.2021.9613682
3. Nunes, P., Santos, J., Rocha, E.: Challenges in predictive maintenance–a review. CIRP J. Manuf. Sci. Technol. **40**, 53–67 (2023)
4. Berghout, T., Benbouzid, M.: A systematic guide for predicting remaining useful life with machine learning. Electronics **11**(7), 1125 (2022)
5. Montero Jimenez, J., Schwartz, S., Vingerhoeds, R.A., Grabot, B., Salaün, M.: Towards multi-model approaches to predictive maintenance: a systematic literature survey on diagnostics and prognostics. J. Manuf. Syst. **56**, 539–557 (2020)
6. Cardoso, D., Ferreira, L.: Application of predictive maintenance concepts using artificial intelligence tools. Appl. Sci. **11**(1), 18 (2021)
7. Jagatheesaperumal, S.K., Rahouti, M., Ahmad, K., Al-Fuqaha, A.I., Guizani, M.: The duo of artificial intelligence and big data for industry 4.0: review of applications, techniques, challenges, and future research directions. https://arxiv.org/abs/2104.02425 (2021)
8. Meng, H., Li, Y.-F.: A review on prognostics and health management (PHM) methods of lithium-ion batteries. Renew. Sustain. Energy Rev. **116**, 109405 (2019)
9. Pater, I., Reijns, A., Mitici, M.: Alarm-based predictive maintenance scheduling for aircraft engines with imperfect remaining useful life prognostics. Reliab. Eng. Syst. Saf.. Eng. Syst. Saf. **221**, 108341 (2022)
10. Maulana, F., Starr, A., Ompusunggu, A.P.: Explainable data-driven method combined with bayesian filtering for remaining useful lifetime prediction of aircraft engines using nasa cmapss datasets. Machines **11**(2), 163 (2023)
11. Cao, Q., Zanni-Merk, C., Samet, A., Reich, C., Beuvron, F., Beckmann, A., Giannetti, C.: Kspmi: A knowledge-based system for predictive maintenance in industry 4.0. Robot. Comput. Integr. Manuf. **74**, 102281 (2022). https://doi.org/10.1016/j.rcim.2021.102281
12. Hagmeyer, S., Zeiler, P., Huber, M.: On the integration of fundamental knowledge about degradation processes into data-driven diagnostics and prognostics using theory-guided data science. PHM Soc. Eur. Conf. **7**, 156–165 (2022). https://doi.org/10.36001/phme.2022.v7i1.3352
13. Sateesh Babu, G., Zhao, P., Li, X.-L.: Deep convolutional neural network based regression approach for estimation of remaining useful life. In: Database Systems for Advanced Applications: 21st International Conference, DASFAA 2016, Dallas, TX, USA, April 16–19, 2016, Proceedings, Part I 21, pp. 214–228 (2016)
14. Zheng, S., Ristovski, K., Farahat, A., Gupta, C.: Long short-term memory network for remaining useful life estimation. In: 2017 IEEE International Conference on Prognostics and Health Management (ICPHM), pp. 88–95 (2017)
15. Wang, J., Wen, G., Yang, S., Liu, Y.: Remaining useful life estimation in prognostics using deep bidirectional lstm neural network. In: 2018 Prognostics and System Health Management Conference (PHM-Chongqing), pp. 1037–1042 (2018)
16. Li, J., Li, X., He, D.: A directed acyclic graph network combined with CNN and LSTM for remaining useful life prediction. IEEE Access **7**, 75464–75475 (2019)
17. Kong, Z., Cui, Y., Xia, Z., Lv, H.: Convolution and long short-term memory hybrid deep neural networks for remaining useful life prognostics. Appl. Sci. **9**(19), 4156 (2019)
18. Mo, H., Lucca, F., Malacarne, J., Iacca, G.: Multi-head cnn-lstm with prediction error analysis for remaining useful life prediction. In: 2020 27th Conference of Open Innovations Association (FRUCT), pp. 164–171. IEEE (2020)

19. Liu, H., Liu, Z., Jia, W., Lin, X.: Remaining useful life prediction using a novel feature-attention-based end-to-end approach. IEEE Trans. Ind. Inform. **17**(2), 1197–1207 (2020)
20. Kumar, K.D.: Remaining useful life prediction of aircraft engines using hybrid model based on artificial intelligence techniques. In: 2021 IEEE International Conference on Prognostics and Health Management (ICPHM), pp. 1–10. IEEE (2021)
21. Yu, K., Wang, D., Li, H.: A prediction model for remaining useful life of turbofan engines by fusing broad learning system and temporal convolutional network. In: 2021 8th International Conference on Information, Cybernetics, and Computational Social Systems (ICCSS), pp. 137–142. IEEE (2021)
22. Liu, Y., Zhang, X., Guo, W., Bian, H., He, Y., Liu, Z.: Prediction of remaining useful life of turbofan engine based on optimized model. In: 2021 IEEE 20th International Conference on Trust, Security and Privacy in Computing and Communications (TrustCom), pp. 1473–1477. IEEE (2021)
23. Asif, O., Haider, S.A., Naqvi, S.R., Zaki, J.F., Kwak, K.-S., Islam, S.R.: A deep learning model for remaining useful life prediction of aircraft turbofan engine on C-MAPSS dataset. IEEE Access **10**, 95425–95440 (2022)
24. Hu, Q., Zhao, Y., Wang, Y., Peng, P., Ren, L.: Remaining useful life estimation in prognostics using deep reinforcement learning. IEEE Access **11**, 32919–32934 (2023). https://doi.org/10.1109/ACCESS.2023.3263196
25. Hong, C.W., Lee, C., Lee, K., Ko, M.-S., Kim, D.E., Hur, K.: Remaining useful life prognosis for turbofan engine using explainable deep neural networks with dimensionality reduction. Sensors **20**(22), 6626 (2020)
26. Peng, C., Chen, Y., Chen, Q., Tang, Z., Li, L., Gui, W.: A remaining useful life prognosis of turbofan engine using temporal and spatial feature fusion. Sensors **21**(2), 418 (2021)
27. Li, Q., Yang, Y.: Transformer model for remaining useful life prediction of aeroengine. In: Journal of Physics: Conference Series, vol. 2171, p. 012072. IOP Publishing (2022)
28. Saxena, A., Goebel, K., Simon, D., Eklund, N.: Damage propagation modeling for aircraft engine runto-failure simulation. In: 2008 International Conference on Prognostics and Health Management, pp. 1–9. IEEE (2008)
29. Idrissi, B.Y., Arjovsky, M., Pezeshki, M., Lopez-Paz, D.: Simple data balancing achieves competitive worst-group-accuracy. In: Conference on Causal Learning and Reasoning, pp. 336–351. PMLR (2022)
30. Sun, J., Li, H., Fujita, H., Fu, B., Ai, W.: Class-imbalanced dynamic financial distress prediction based on adaboost-svm ensemble combined with smote and time weighting. Inf. Fusion **54**, 128–144 (2020)
31. Fotouhi, S., Asadi, S., Kattan, M.W.: A comprehensive data level analysis for cancer diagnosis on imbalanced data. J. Biomed. Inform. **90**, 103089 (2019)
32. Duan, M., Liu, D., Chen, X., Liu, R., Tan, Y., Liang, L.: Self-balancing federated learning with global imbalanced data in mobile systems. IEEE Trans. Parallel Distrib. Syst. Distrib. Syst. **32**(1), 59–71 (2020)
33. Cheung, R.C., Lee, D.-U., Luk, W., Villasenor, J.D.: Hardware generation of arbitrary random number distributions from uniform distributions via the inversion method. IEEE Trans. Very Large Scale Integr. (VLSI) Syst. **15**(8), 952–962 (2007)
34. Patel, S.K., Surve, J., Katkar, V., Parmar, J., Al-Zahrani, F.A., Ahmed, K., Bui, F.M.: Encoding and tuning of THz metasurface-based refractive index sensor with behavior prediction using xgboost regressor. IEEE Access **10**, 24797–24814 (2022). https://doi.org/10.1109/ACCESS.2022.3154386
35. Hodson, T.O.: Root-mean-square error (RMSE) or mean absolute error (MAE): when to use them or not. Geosci. Model Dev. **15**(14), 5481–5487 (2022)
36. Azadkia, M.: Optimal choice of k for k-nearest neighbor regression. arXiv: Statistics Theory (2019)

HTTP Flood DOS Attack Detection on Big Data Using Data Mining

Mohamed Haddadi[1(✉)], Abdelhamid Khiat[2], Yasmina Abidi[3], and Yaakoub Derradji[3]

[1] Department of Commercial Sciences, Faculty of Economic, Commercial, and Management Sciences, M'hamed BOUGARA University of Boumerdes, Independence Avenue, 35000 Boumerdes, Algeria
m.haddadi@univ-boumerdes.dz

[2] Networks and Distributed Systems Division Research Center on Scientic, and Technical Information, Ben Aknoun, Algeria
a.khiat@dtri.cerist.dz

[3] Department of Computer Science, Faculty of Sciences, University of Ferhat Abbas Setif1, Setif, Algeria

Abstract. Nowadays, increasing use of Internet connection, security becomes a huge challenge for individuals as well as governments and organizations. Therefore, in the last decade, the world is moving towards green computing in the purpose either to store energy or to decrease operational costs. So, this new technology uses web servers to provide web applications to end user. Generally, these web servers become unavailable because of HTTP flood DoS attack. This paper applied data mining techniques such as WEKA and TANAGRA tools for data preprocessing in the aim to get a consistent data in one hand, and for classifying the traffic in normal behavior or anomaly using like J48, Random tree, SMO, Naïve bayes, IBK, and combined classifier in the other hand. For this aim, a well-known big data set is used which called http_csic_2010_full dataset. Results show that the two data mining tools are well close with all the classifiers, just MLP in WEKA, and CVM in TANAGRA they didn't perform well in terms of detection accuracy.

Keywords: Data preprocessing · Weka tool · Tanagra tool · HTTP flood DoS attack · http_csic_2010_full dataset

1 Introduction

Nowadays, the Internet is widely used in organizations as well as individuals everywhere because of its significant benefits like online services, e-commerce, e-banking, and so on. However, in the last decades, the number of attacks (hackers or threats) is increasing seriously in parallel with the increasing number of users using Internet connection day by day. Hence, a web security threat or issue is defined as a important malicious activity that exclusively targets one or multiple components of web application's architecture such as the user's browser or the web application hosting server [1]. So, the aim of these threats is to deny legitimate users to access services delivered by Internet itself.

M. R. Laouar et al. (Eds.): ICISAT 2023, LNNS 981, pp. 37–49, 2024.
https://doi.org/10.1007/978-3-031-60591-8_4

Currently, one of most hard attack is Denial of Service (DoS) and its variants such as Distributed Denial of Service (DDoS) attack which can interrupt the availability of applications delivered by web servers. So, this attack can occur at many layers of the OSI (Open System Interconnection) model especially the application layer.

In fact, this attack can be generated using high rate and low rate against one or more web servers [2]. For instance, high (low) rate DoS attack is mitigated because of its high (low) traffic generated which is different to that generated by authorized users. In parallel, many proposed approaches are applied in network, and transport layers using different techniques such as that mentioned in [3]. But, in application layer, HTTP flood DoS attack remains a huge challenge and it's the most known attacks which is a type of volumetric DDoS attack designed to overwhelm a web server with HTTP requests. Once the web server is saturated with requests and is unable to respond to normal traffic [4]. In fact, this attack becomes more and more hard to detect especially with low rate because its generated traffic is similair to that generated by legitimate users in order to deny the authorized users to acces sevices provided by web servers. Moreover, this type of attack is also called slow HTTP DoS attack [5] so that it also inceases the capacity of web servers on CPU and memory.

Generally, HTTP Flood DoS attacks are mostly classified into two main classes: http-GET and http-POST flood [5]. But, in the great majority of cases, the attackers use HTTP-Get DoS attack which will be focused in this paper. Also, in this attack, the attacker sends HTTP-Get requests slowly to web server to keep the server busy waiting for the rest of the request.

Because of the calmness of the problem various defense mechanisms have been applied to detect these attacks. The purpose of our subject is to detect HTTP flood DoS attacks using Data Mining (DM) which has emerged as a significant technology for gaining knowledge from vast quantities of data. The goal of most data mining approaches is to develop generalized knowledge, rather than identify information about specific individuals [6]. Moreover, DM is a process of extraction of useful information and patterns from huge data. It is also called a knowledge discovery process, knowledge mining from data, knowledge extraction, or data/pattern analysis [7].

Currently, we have proposed an approach in which we use data mining classification techniques to detect http flood DoS attacks using WEKA and TANAGRA tools. The approach is based on the classification of 'normal' requests against 'anomalous' requests in the sense of a DoS attack. The aim is to investigate and evaluates the performance of the Decision Tree, Multilayer Perceptron, Support Vector Machine, Naive Bayes, K-Nearest Neighbours, and Random Tree algorithms to detect DoS attacks, and compares them with each other. These algorithms were tested with http_csic_2010_full.arff dataset [8].

The remainder of this pape is divided as follows: Sect. 2 outlines related work whereas, Sect. 3 presents our used data set. Implementation and results discussion are done in Sect. 4. Finally, A brief conclusion is given in Sect. 5.

2 Related Work

This section gives mutiple existing approaches, available in recent literature, that detect DoS attack, especially HTTP flood attack at application, and UDP-flood attack at network, and transport layers. Each of which comes with its advantages and shotcomings. Mainely, these techniques are judged in terms of a few important performance metrics such as detection accuracy. Therefore, this related work aims to bring the latest proposed algorithms against HTTP flood attack and its variants.

We begin with the proposal of Mohammad et al. [9], who proposed an analysis of the KDD Cup 99 dataset using K-means clustering-based data mining which is an unsupervised learning technique. This dataset has four types of attacks Denial of Service (DoS), User to Root (U2R), Remote to Local (R2L), and Probing (PROBE). So, it consists of 41 attributes which are 10% of the original dataset means 500,000 rows (records). Each of these has a label as normal or attack. In this dataset, the training set contains 19,69% normal and 80,31% attack connections in three protocol types such as TCP (Transmission Control Protocol), UDP (User Datagram Protocol), and ICMP (Internet Control Protocol). The results show that the proposal increases the accuracy detection rate for suspicious activity, and improves the efficiency of the real intrusion detection model.

Another technique is developed by Mouhammed et al. [10], who applied Three machine learning Algorithms like MLP (Multilayer Perceptron), Random Forest, and Naïve Bayes to classify the DDoS types of attack namely: Smurf, UDP-Flood, and HTTP-Flood using new dataset which contains five classes. Each class has 27 features extracted from application and network layers. Experiments show that The MLP classifier achieves the highest accuracy rate of 98.63%.

Beghdad et al. [11], who enhanced two well-known algorithms ID3 (Iterative Dichotomiser 3), and MLP that are used for intrusion detections. The enhancement is based on the use of 14 appropriate attributes of the KDD dataset for the training and testing phases of the two presented solutions because some attributes are not useful for training artificial approaches. For implementing both algorithms (ID3 and MLP), they have used the Java language and WEKA tool. The simulation results show that (ID3) has the highest detection accuracy of 99.88%, and MLP seems clear to have 99.10%.

Yogita et al. [12], who proposed a method of intrusion detection using SVM (Support Vector Machine) and Gaussian RBF (Radial Basis Function) kernel. NSL-KDD Cup'99 dataset is conducted with 10 folds cross-validation and Gaussian RBF kernel of SVM. Results show that the method achieves a detection accuracy of 98.5749%.

Moreover another approach is proposed by Waguih et al. [13], to detect DoS attacks using data mining classification techniques. This approach is based on J48 (C4.5) decision tree algorithm. A benchmark dataset called 1998 DARPA is conducted to evaluate the approach. Experiments show that the approach achieves an accuracy rate of 99%.

Anand et al. [14], who applied Data Mining techniques such as Naive Bayes, K-Means, SVM, CART Random Forest, K-Means to detect DoS attacks so that NSL-KDD dataset is conducted to compute the detection accuracy for each of which. Results show that the performance of Naive Bayes and simple K-Means is very poor, whereas SVM performance is fairly good. But, Decision Tree(CART), and Random Forest achieved a high detection accuracy.

Deepak et al. [15], presented an ontology based on the intrusion detection framework for the detection of DoS at the application level HTTP protocol. GoldenEye DoS tool dataset is conducted to test the proposal using semantic rules. Experiments show that the system provides early detection of HTTP flood DoS attacks in two seconds and an improved detection rate using a time window threshold mechanism in the semantic rule. The system also achieves a higher detection rate of 94.89% without a threshold in the semantic rule to detect DoS attack at the application level HTTP protocol.

Alauthman et al. [16], who suggested different approaches to application-layer Dos attack detection. Therefore, they tested Random forest, SVM, Logistic Regression, and Gaussian NB to be malicious activity detectors. Results show that Random forest achieves around 99.0% detection rate.

Hao et al. [17], used Naive Bayes algorithm, artificial neural network, and logistic regression to detect DoS attacks. Experiments indicated that the artificial neural network performs better than Naive Bayes and logistic regression.

3 HTTP_csis_2010_full Dataset

We have used **http_csic_2010_full** dataset [8] which is one of the mostly used recent benchmark dataset at application layer in comparison of other dataset which are out of date such as KDD Cup 1999 dataset [18] which is used only for cybersecurity in network and transport layers. In addition to that, this dataset was created by Peter Scully PhD in 2010. In fact, it has a capacity of ~ 99.3 MB *(uncompressed) with* 223585 instances. Each of which has 18 attributes (features (columns)) which are as follows: *"index", "method", "url", "protocol", "userAgent", "pragma", "cacheControl", "accept", "acceptEncoding", "acceptCharset", "acceptLanguage", "host", "connection", "contentLength", "contentType", "cookie", "payload", "label". In the label attribute the instances are labeled with two attributes (e.g.,* norm or anom) as shown in Table 1.

Table 1. Example of description of two records

Record description	Label
0,"GET","http://localhost:8080/tienda1/publico/anadir.jsp","HTTP/1.1","Mozilla/5.0 (compatible Konqueror/3.5 q = 0.9,text/plain q = 0.8,image/png,*/* q = 0.5","x-gzip, x-deflate, gzip, deflate","utf-8, utf-8 q = 0.5, * q = 0.5","en","localhost:8080","close","null","null","JSESSIONID = B92A8B48B9008CD29F622A994E0F650D","id = 2","anom"	HTTP DoS Attack Request
2895,"GET","http://localhost:8080/tienda1/global/creditos.jsp","HTTP/1.1","Mozilla/5.0 (compatible Konqueror/3.5 Linux) KHTML/3.5.8 (like Gecko)","no-cache","no-cache","text/xml,application/xml,application/xhtml + xml,text/html q = 0.9,text/plain q = 0.8,image/png,*/* q = 0.5","x-gzip, x-deflate, gzip, deflate","utf-8, utf-8 q = 0.5, * q = 0.5","en","localhost:8080","close","null","null","JSESSIONID = E10C52B8A9DBF6D251083742C2385B1D","""norm"	Normal Request

4 Implementation

4.1 Weka Tool

WEKA tool supports several standard data mining techniques like data pre-processing, clustering, classification, regression, visualization, and feature selection [19]. So, these techniques are used to carry out a classification experiments using the http_csic_2010_full dataset. We used WEKA tool to evaluate the standard techniques with the same values and parameters such as the accuracy of each used data mining technique such as J48, Random tree, SMO, Naïve bayes, IBK, and combined algorithm (J48/Random Tree) using four cases such as Percentage split, Cross-validation, Supplied test set, and training set.

A. WEKA Metrics

WEKA tool is used to carry out all experiences. Http_csic_2010_full dataset is conducted to compare the performance a few possible WEKA classifiers using confusion matrix which is used to describe the performance of a classification model:

- True Positive (**TP**): The classifier predicts normal behavior as normal.
- True Negative (**TN**): The classifier predicts anomaly behaviors as anomalous.
- False Positive (**FP**): The classifier predicts TRUE, but the correct class was FALSE (request is normal)
- False Negative (**FN**): the classifier predicts FALSE (request is normal), but are TRUE (anomalous).

The accuracy of a model depends on its quality, the number of training samples, and the model parameters. Accuracy is calculated as the total of two correct predictions (TP + TN) divided by the total number of data sets (TP + FP + TN + FN). Accuracy = (TP + TN) / (TP + FP + TN + FN).

The best accuracy is 1.0 and the worst is 0.0.

B. WEKA Cleaning Data

This step is the most important for creating a consistent dataset by following the following important operations as follows:

- Identifying and Handling the Missing Values
  The first step is to import the dataset using many methods such as WEKA GUI Chooser (WEKA V3.9.5). Then in the step for identifying and handling the missing values using several techniques manually or automaticaaly so that for For the categorical values we fill in the most common noun, and for the numerical fill in the previous number, or calculate the average. One of the most used way to handle missing values in WEKA using ArffViewer is presented as follows:

- Loading dataset in Explorer
- Click on Choose/WEKA/Filters, unsupervised, and attribute.
- Click on ReplaceWithMissingValues, and save the dataset
- Return to Weka GUI Chooser, click on Tools

- And ArffVIEWER
- The ARFF-Viewer window appears
- Click on File/Open, a new window appears « Ouvrir», open the new dataset where the missing values are appear
- Double click on the case where there is no value, and we write or chose a new value. We do this manually with all fields blank. This method is very time consuming and tiring as shown in Fig. 1.

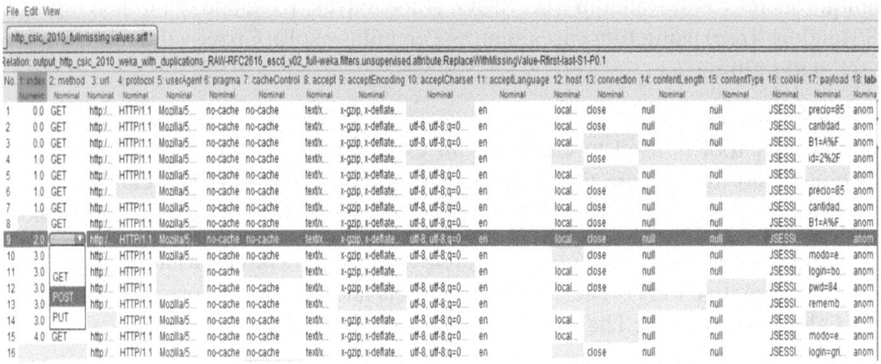

Fig. 1. Handling missing values manually using WEKA

- Smooth the Noisy Data

  Noisy generally means random error or containing unnecessary data points. The incorrect values cause data entree problems, data transmission problems, etc. Here are some of the methods to handle noisy data [20]. One of the mostly used way to handle NOISY DATA in WEKA is shown as follows:
- **Combining between computer and human:**

  combining computer and human inspection by detecting noisy data and handling it. We use a filter to smooth this noisy data based on supposed rules according to human inspection:
- The values more than 100% or less than 0% are errors (noisy).
- The values between 45% and 49% are critical and must become 50%.

Here we use WEKA for this purpose by following these steps:

1- Loading the dataset

2- Click on Choose/WEKA/Filters, unsupervised, and attribute.

3- Click on NumericCleaner

4- Change the settings as follows and as shown in Fig. 2.

For the first rule:

- maxDefault: 100
- maxThreshold: 100
- minDefault: 0
- minThreshold: 0

For the second rule:

- CloseTo: 47
- closeToDefault: 50
- clmoseToTolerance: 3

AttributeIndices: write the number of the attribute which we need to handle its values. Click on apply.

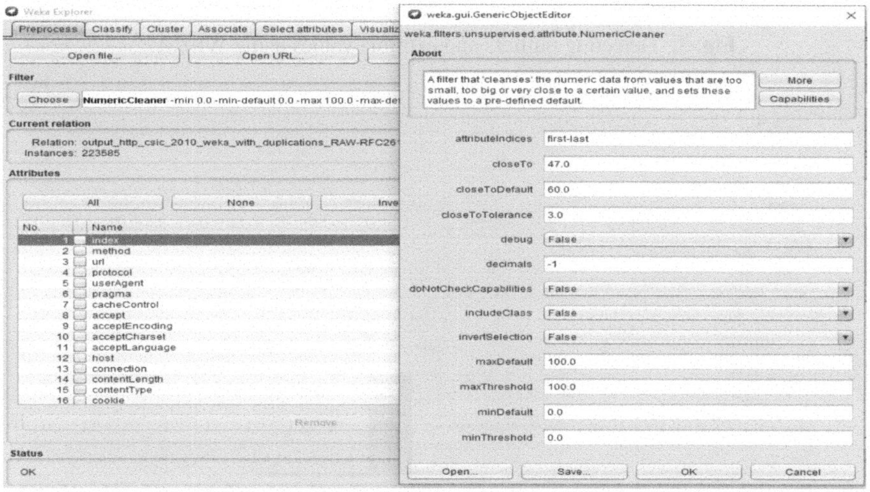

Fig. 2. Change settings for NumeralCleaner filter using WEKA

- Remove Outliers
  Outliers have not complied with the general behavior (database). Exactly, outliers are data point that is significantly different from other data points in a data set. In WEKA, we remove outliers in this way:
- Loading dataset in Explorer
- Click on Choose/WEKA/Filters, unsupervised, and attribute.
- Click on InterquartileRange,
- Click on apply. Figure 3 shows how the outliers and extreme values are detected:

How to move outliers and extreme values in WEKA:

1- Click on Choose/WEKA/Filters, unsupervised, and instance.
2- Click on RemoveWithValues,
3- Change the settings as shown in Fig. 4.
4- Click OK, and apply.

- Remove Useless Attributes
  Remove useless attributes that do not vary at all, or vary too much. In WEKA, we remove outliers in this way:

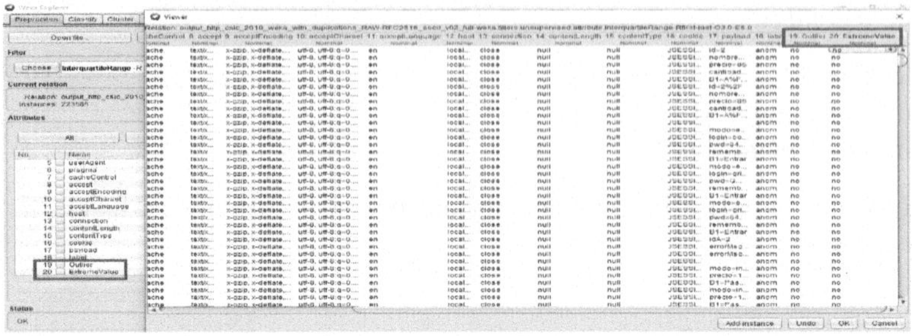

Fig. 3. Detecting outliers and extreme values using WEKA.

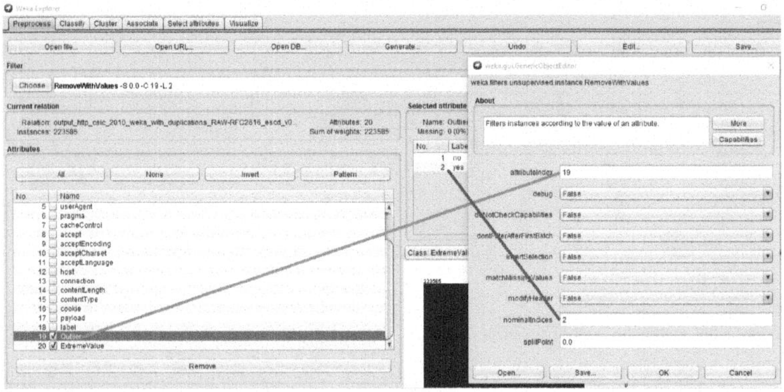

Fig. 4. Changing the setting to remove outliers and missing values using WEKA

- Loading dataset in Explorer
- Click on Choose/WEKA/Filters, unsupervised, and attribute.
- Click on RemoveUseless,
- Click on apply. Figure 5 shows how the outliers and extreme values are detected:

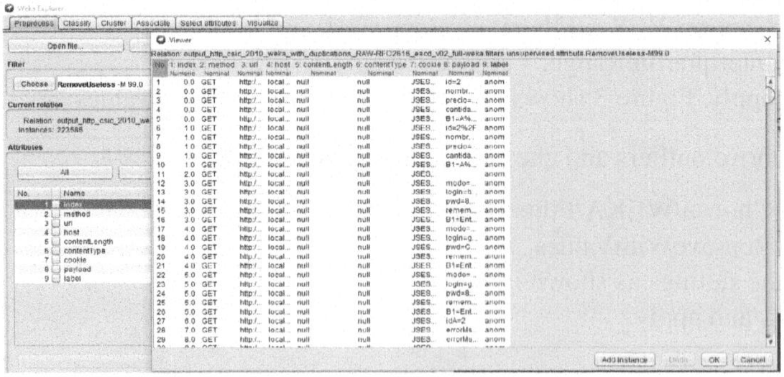

Fig. 5. Remove useless attributes using WEKA

4.2 Results Discussion

In this section, we applied three cases such as Training set, percentage split and 10-fold cross-validation on http_csic_2010_full dataset as follows:

- Training set (or with supplied test set)
  Here, we use training set (use training set in "test option" which means that test on the same set that the classifier is trained on (dataset is splitting in training and testing sets), where, supplied test set (this case gives the same result as the first one) (test on a specific dataset).
  In Table 2, we compare the detection accuracy of a few classifiers such as J48, Random tree, SMO, Naïve bayes, IBK, and combined algorithm (J48/Random Tree) in terms of detection accuracy. For this aim, http_csic_2010_full dataset is conducted to compute the detection accuracy of each of which.

Table 2. WEKA classifiers comparison using training set

Detection Accuracy							
Classifiers Test options	J48	Random tree	SMO	Naive Bayes	MLP	IBK	J48 + Random Tree
training set (or with Supplied test set)	100%	100%	100%	100%	62.74%	100%	100%

According to the Table 2, J48, Random tree, SMO, Naïve bayes, IBK; and combined classifier (J48/Random Tree) seem to have the highest detection accuracy of 100% using training set, whereas MLP seems to have the lowest detection accuracy of 62.7412%. This comparison is done under similar features of the previous dataset.

- Percentage Split
  In this case, we use percentage split which means that train on the percentage of the dataset, and test on the remainder.
  In the following, we compare the detection accuracy of a few WEKA classifiers such as J48, Random tree, SMO, Naïve bayes, IBK; and combined algorithm (J48/Random Tree) as illustrated in Table 3. Using the similar and the same well-known benchmark dataset.

In Table 3, J48, Random tree, SMO, Naïve bayes, IBK; and combined classifier (J48/Random Tree) seem to have the highest detection accuracy of 100%, whereas MLP seems to have the lowest detection accuracy of 53.4787%. This comparison is done under similar features of the previous dataset called http_csic_2010_full dataset.

- 10-fold cross validation
  In 10-fold cross validation, we arbitrarily split the samples into 10 distinguish subsamples: 6 of which are used for training and the remaining part of subsamples is used for testing.

Table 3. WEKA classifiers comparison using percentage split

Detection Accuracy							
Algorithms Test options	J48	Random tree	SMO	Naive Bayes	MLP	IBK	J48 + Random Tree
Percentage split	100%	100%	100%	100%	53.47%	100%	100%

In Table 3, we compare the detection accuracy of a few classifiers such as J48, Random tree, SMO, Naïve bayes, IBK; and combined algorithm (J48/Random Tree) using the similar and the same well-known benchmark dataset called http_csic_2010_full dataset.

Detection Accuracy							
Algorithms Test options	J48	Random tree	SMO	Naive Bayes	MLP	IBK	J48 + Random Tree
10-Cross-validation	100%	100%	100%	100%	46.51%	100%	100%

According to the Table 3, J48, Random tree, SMO, Naïve bayes, IBK; and combined classifier (J48/Random Tree) seem to have the highest detection accuracy of 100% using 10-Cross-validation, whereas MLP seems to have the lowest detection accuracy of 46.5147%. This comparison is done under similar features of the previous dataset called http_csic_2010_full dataset.

- **Global comparison usnig the three previous cases**
  Here, we take into account the three cases as training set (or with supplied test set), 10-Cross-validation and percentage split, and the all used classifiers already cited like J48, Random tree, SMO, Naïve bayes, IBK; and combined algorithm (J48/Random Tree) as shown in Fig. 2.

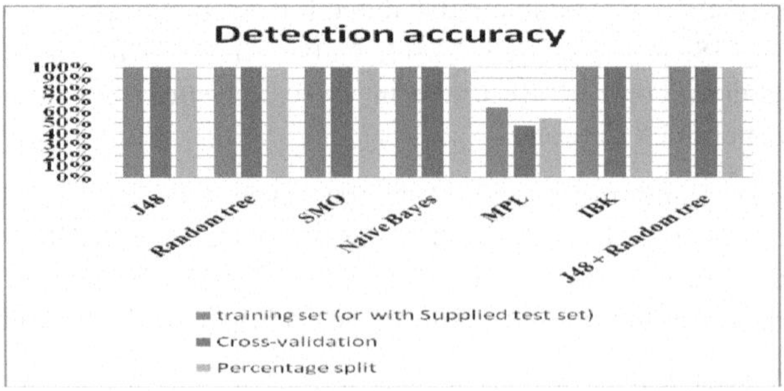

Fig. 6. Global comparisons using the three previous cases

According to the Fig. 6, J48, Random tree, SMO, Naïve bayes, IBK; and combined algorithm (J48/ Random Tree) seem to have the highest detection accuracy of 100% in terms of training set (or with Supplied test set), Cross-validation, and Percentage split, whereas MLP seems to have the lowest detection accuracy of 62.7412% using training set (or with supplied test set), 46.5147% using Cross-validation, and 53.4787% using percentage split. This comparison is done under similair features of the previous dataset called http_csic_2010_full dataset.

4.3 TANAGRA Tool

In TANAGRA, the accuracy is calculated from the confusion matrix applying the following equation:

$$\text{Detection accuracy} = (\text{TP} + \text{TN})/(\text{TP} + \text{FP} + \text{TN} + \text{FN}).$$

In Table 4, Naive Bayes has the highest accuracy (100%) with C4.5, MLP, and Rnd Tree, and the two combining algorithms (C4.5, and Rnd Tree). But the lowest accuracy is for CVM (47.3657%) because of the largest dataset, and the hyper parameters are not chosen well.

Table 4. Detection accuracy of some TANAGRA classifiers

Algorithms	Accuracy
Naive Bayes	100%
Random Tree	100%
C4.5	100%
K-NN	Not applied
MLP	100%
CVM	47.3657%
Bagging (C4.5/Rnd Tree)	100%

5 Conclusion

In this work, we applied different supervised algorithms such as Decision tree, Random Tree, Naive Bayes, MLP, SVM, and others to classify HTTP requests as normal or anomalous (attack). We implemented these algorithms using WEKA and TANAGRA tools in the purpose to search the highest accuracy, and the best classifier to classify HTTP requests. For this aim, CIC-IDS2017 dataset test tool is used to preprocess data and it is considered such as a big data. We applied data preprocessing in the purpose to give a consistent data for these classifiers so that we applied supervised data mining techniques for this objective such as preprocessing tasks like Data cleaning such as handling duplicate data, handling irrelevant data, handling structural errors, handling missing values (categorical data & numerical data), handling outliers values. Then, we applied the previous cited classifiers on the consistent dataset. The results show that WEKA has the best detection accuracy. So, the algorithms performed well with WEKA tool. From the results, we saw that the two data mining tools are well close with all the classifiers, just MLP in WEKA, and CVM in TANAGRA they didn't perform well.

6 Future Work

In the future, the performance of the proposed model has to be improved and it has to be implemented in a real application layer. The authors would have to get permission from Internet providers.

Acknowledgment. This research received no specific grant from any funding agency in the public, commercial, or notfor-profit sectors.

References

1. Sadqi, Y., Maleh, Y.: A systematic review and taxonomy of web applications threats. Inf. Sec. J. A Global Persp. **31**(1), 1–27 (2022). https://doi.org/10.1080/19393555.2020.1853855
2. Haddadi, M., Beghdad, R.: DoS-DDoS: taxonomies of attacks, countermeasures, and well-known defense mechanisms in cloud environment. Edpacs **57**(5), 1–26 (2018). https://doi.org/10.1080/07366981.2018.1453101
3. Mohamed, H., Ouldmohamed, Y., Nacera, B.: A single-packet IP TraceBack: combating dos-ddos attacks. EDPACS **66**(4), 1–12 (2021). https://doi.org/10.1080/07366981.2021.1999800
4. Cloudflare. (n.d). HTTP Flood DDoS Attack. Retrieved November 15, 2021, from https://www.cloudflare.com/learning/ddos/http-flood-ddos-attack/
5. Kshirsagar, D., and Kumar, S.: HTTP flood attack detection using ontology. In: Proceedings of the International Conference on Advances in Information Communication Technology & Computing, pp 1-4. (2016). https://doi.org/10.1145/2979779.2979794
6. Vaidya, J., Zhu, Y.M., Clifton, C.W.: Privacy and Data Mining, vol. 3, LNCS, pp. 1–5. Springer, New York (2006). https://doi.org/10.1007/978-0-387-29489-6
7. Raval, K.M.: Data mining techniques. Int. J. Adv. Res. Comput. Sci. Softw. Eng. **2**(10), 13–30 (2012)
8. https://petescully.co.uk/research/csic-2010-http-dataset-in-csv-format-for-weka-analysis/. accessed on May 19th, (2022)
9. Siddiqui, M.K., Naahid, S.: Analysis of KDD CUP 99 dataset using clustering based data mining. Int. J. Datab. Theory Appl. **6**(5), 23–34 (2013)
10. Alkasassbeh, M., Al-Naymat, G., Hassanat, A., Almseidin, M.: Detecting distributed denial of service attacks using data mining techniques. Int. J. Adv. Comput. Sci. Appl. **7**(1), 436–445 (2016)
11. Beghdad, R., Bechar, K., Bouali, M., Haddadi, M.: Neural networks and decision trees for intrusion detections: enhancing detection accuracy. EasyChair, (2897) (2020)
12. Bhavsar, Y.B., Waghmare, K.C.: Intrusion detection system using data mining technique: support vector machine. Int. J. Emerg. Technol. Adv. Eng. **3**(3), 581–586 (2013)

13. Waguih, H.: A data mining approach for the detection of denial of service attack. IAES Int. J. Artif. Intell. **2**(2), 99 (2013)
14. Keshri, A., Singh, S., Agarwal, M., Nandiy, S. K.: DoS attacks prevention using IDS and data mining. In: 2016 International Conference on Accessibility to Digital World (ICADW) (pp. 87–92). IEEE (2016)
15. Kshirsagar, D., Kumar, S.: An ontology approach for proactive detection of HTTP flood DoS attack. Int. J. Syst. Assur. Eng. Manag. (2021). https://doi.org/10.1007/s13198-021-01170-3
16. Alauthman, M., and Albesani, G.: Application-Layer Dos Attack Detection Using Machine Learning. Proceedings Book, vol 58, (2019)
17. Li, Z., Zhang, H., Shahriar, H., Lo, D., Qian, K., Whitman, M., and Wu, F.: Denial of service (dos) attack detection: Performance comparison of supervised machine learning algorithms. In 2020 IEEE Intl Conf on Dependable, Autonomic and Secure Computing, Intl Conf on Pervasive Intelligence and Computing, Intl Conf on Cloud and Big Data Computing, Intl Conf on Cyber Science and Technology Congress (DASC/PiCom/CBDCom/CyberSciTech) (pp. 469–474) (2020), IEEE
18. KDD cup 99 dataset : http://kdd.ics.uci.edu/databases/kddcup99/kddcup99.html accessed on 12/12/2021
19. https://www.cs.waikato.ac.nz/ml/weka/. accessed on Mar. 30th, 2022)
20. Witten, I.H., Frank, E.: Data mining: practical machine learning tools and techniques with Java implementations. ACM SIGMOD Rec. **31**(1), 76–77 (2002)

Towards an Automatic Scoring for Code Assessment in the Universities: A Case Study the Algerian University

Karima Boussaha[1(✉)] and Samia Drissi[2]

[1] Department of Mathematics and Computer Science, LIAOA Laboratory, University of Oum El Bouaghi, Oum El Bouaghi, Algeria
karima.boussaha@univ-oeb.dz

[2] Department of Mathematics and Computer Science, LIM Laboratory, Mohamed-Cherif Messaadia University, Souk Ahras, Algeria
s.drici@univ-soukahras.dz

Abstract. The assessment of learners, whether conducted in traditional classrooms or in the Computer Environment for Human Learning (CEHL), has consistently been a subject of ambiguity and lack of clarity for both the evaluated and the evaluator on the one hand, and between the evaluators themselves on the other hand. The approach taken during the validation of the results of the automated assessment tools was to make an analytical comparison with the scores given by the human experts. The goal of this research paper is to compare and contrast two versions of the S_Onto_ALPPWA automated assessment scoring tool with the expert's scoring. To achieve the underlined goal, we used the correlational research methodology to demonstrate the extent of the correlation between S_Onto_ALPPWA and the manual scores provided by the experts. Our obtained results indicate that, in the majority of instances, the S_Onto_ALPPWA scores are closely aligned with the experts' scores. In our research, we used Ward's hierarchical clustering techniques to categorize the experts into three distinct groups: lenient, severe, and moderate experts. This approach was employed to gather relevant information.

Keywords: Computer Education · Programming Assessment of languages · Automated assessment graders · Learning programming · learners' assessment · practical works activities · learners' assessment · Code assessment · Human scoring · Automatic scoring · ontologies · Automatic feedback

1 Introduction

Within the academic literature, there exist two distinct approaches for evaluating students in basic programming courses: paper-and-pencil testing (PPT) and computer-based assessment (CBA) [1]. The use of pen-and-paper testing is often impractical because of the labor costs associated with it. As a result, teachers often face the challenge of assessing a high volume of assignments. This makes the assessment process demanding, routine, and tiring for teachers due to the complex nature of evaluating student

M. R. Laouar et al. (Eds.): ICISAT 2023, LNNS 981, pp. 50–60, 2024.
https://doi.org/10.1007/978-3-031-60591-8_5

programs. Two possible approaches to solve the problem at hand: augmenting the teaching workforce or using computer-based assessment (CBA) technologies. This particular approach offers several advantages as compared to the first technique. Notable benefits include objective evaluation, elimination of halo effects, a decrease in manual effort, the potential for errors, and marking workload. Kalogeropoulos et al. [1] conducted a study on. The article provides an analysis of the distinctions between the two evaluation methodologies.

Teaching or studying programming at academic institutions poses significant challenges for both professors specialized in this topic and students, due to its complex and abstract character [3, 4].

When total containment was imposed by many countries, educational activities were suspended, necessitating adapting teaching methodologies to accommodate the new normal that does not depend solely on face-to-face interactions. The distance education method was applied, which raised new challenges, especially when evaluating practical courses that have a large number of students for each teacher. In such cases, the assessment techniques and procedures used in traditional face-to-face teaching may be inadequate and unsuitable for implementation in an entirely distance learning environment. Under these exceptional circumstances, we put into practice the second version of our system developed in 2015 called (S_Onto_ALPPWA) [5] specifically for this purpose to address these problems as a result of its ability and suitability to the completely online assessment environment. This paper presents a case study comparing the traditional approach to student assessment to the completely online format of a programming course in higher education. This shift was forced by the COVID-19 pandemic and facilitated through the use of our automated assessment tool known as S_Onto_ALPPWA [6].

To evaluate the new assessment system. We compare in this study, the score equivalence of the computer-based assessment (CBA) system called S_Onto_ALPPWA [5, 6] and paper-and-pencil testing PPT (teachers) in an introductory computer programming practical works activities. In particular, we investigate the scores awarded by two versions of the developed System [5, 6], and the teachers. The research questions that guided this study were:

1. Which of the two versions of the designed system [5, 6] is capable of providing scores that closely approximate those given by teachers?
2. Is the improved system fully reliable in pandemic conditions for a full online assessment? Is it effective as a result of its ability and suitability for a complete remote evaluation environment?

The rest of the paper is structured as follows. In Sect. 2, we introduce a comprehensive overview of relevant literature. In Sects. 3 and 4, we present a concise overview of the functionality of the developed system in its old and improved versions. Section 5 of the paper provides a comprehensive account of the method design, including a thorough explanation of the gathered data, the presentation of obtained results, and an analysis of the scoring outcomes. Section 6 includes an in-depth discussion of the subject treated. In conclusion, this study was concluded through a deep analysis of the results obtained and a discussion of potential new directions in the research.

2 Related Work

Automated evaluation systems are very important and play a serious role in student learning courses as they facilitate programming tasks and provide quick feedback to help students who have difficulties complete their courses in the best conditions. The primary motivation behind incorporating these systems into programming classes is their ability to significantly and effectively ease teachers' responsibilities while maintaining a high level of accuracy.

Numerous scientific publications have emerged in recent years, focusing on literature evaluations of Automated assessment systems [7–10]. Based on many dimensions, literature studies have categorized these systems, including the kind of evaluation used, which may be automated, semi-automated, or human. According to the authors in [7], the evaluation procedure may be categorized into three main types: student-centered, instructor-centered, or a combination of both (hybrid) [7]. When, the analytical technique can be classified as static, dynamic, or a combination of both (hybrid). According to the authors in [9, 10]. In their study, Keuning et al. [8] undertook a comprehensive analysis of automated assessment tools, specifically examining the feedback produced by these systems. Automated assessment systems of this type are generally classified as formative assessment tools. These tools can be considered summative assessment tools if they only provide scores or percentages. With the aim of influencing and modifying the learner's cognitive processes or behavioral patterns, feedback is often used as a means of conveying information to the learner to enhance the learning process. In this context, we have created an automated assessment system that primarily focuses on providing student-centered feedback.

Another type of review we found during our in-depth analysis of the literature on automatic code assessment. This particular evaluation focuses on the examination of tools that have been created specifically to facilitate automated code assessment [11, 12]. The primary focus of their analysis is based on the different functions that can be achieved, along with the inherent challenges associated with developing them. The authors in the article [13] present three generations of assessment systems. The first-generation group focused mostly on basic analyses and produced summary reports using raters. The second generation consisted of tool-oriented platforms which included testing engines and other tools such as pattern checkers. Finally, the last group of the third generation included web-oriented tools.

Sébastien Combéfis in [14] has done a thorough examination of the review papers to gain a comprehensive understanding of the research on automated code assessment in higher education. From 2005 to 2021, eighteen references were surveyed and reviewed, which included all aspects of automated code evaluation methodologies and tools. As a result of this survey, six primary concerns were identified regarding the development of automated code review systems. These concerns are summarized in the following aspects:

- Determining the specific coding and program elements that should be evaluated.
- Selecting appropriate methods and techniques for code analysis.
- Defining the kinds of feedback that should be provided to both learners and instructors.
- Designing effective systems to facilitate automated code assessment.

- Integrating these systems into the learning process and their application in educational settings.
- Evaluating the quality and impact of the assessments generated through automation.

3 The First Version of the Developed System Overview

We have introduced an initial version of a novel self-assessment environment designed to enable students to examine and verify their performance in Object-Oriented Programming. The proposed initial developed version being presented takes into account both the structural and dynamic features of object-oriented codes. The primary purpose of the developed environment is to assess the students' codes by comparing them with the professor's via a syntactic comparison of their UML graphical representations, namely class and communication diagrams. The developed environment offers to the learner a set of functionalities [6]:

• Firstly, this tool enables the instructor to compose C++ code and subsequently generate UML diagrams from it. These diagrams serve as reference descriptions during the evaluation process.

• Secondly, it also allows students to create their own codes, generate UML diagrams from them, and initiate the evaluation process.

• Lastly, we employ the matching techniques that we have proposed, which yield scores between the student's code and the instructor's. (for more details see the author's publication in [6]).

Taking into account only syntactic comparison is the primary limitation of the first version of the developed system [6]. To overcome this weakness, we included the semantic codes features, namely semantic comparison, in the second version of our development system (Refer to paragraph 4).

4 The Improved Developed System Overview

The improved system aims to facilitate the comparison between students' codes and those created by the instructor. The proposed learning environment is specifically designed for novice learners. This developed system facilitates the evaluation of students' object-oriented codes by providing a platform for comparison with the codes developed by the instructors. The fundamental concept is using ontologies in their visual depictions as a means to indirectly compare object-oriented codes. To provide the student with notes and errors he made in his code. The developed improved system goes through three basic stages [5]:

- Two ontologies are generated based on the learner's code and the professor's code.
- To measure the degrees of scores of similarities between the student's codes and the teacher's, the system utilizes matching algorithms. Based on the presented results, an observation is sent to the learner.
- By providing a comprehensive list of semantic and grammatical errors identified in students' codes as feedback, their performance is assessed in this way. (For more details see the author's publication in [5]).

5 Experimental Study

5.1 Experimental Procedure

We recruited 12 teachers from different institutions in Algeria and conducted experiments across three phases to demonstrate the efficacy of the proposed approach and the developed system. The role of the teacher is to assess the students' proposed solutions to two programming exercises. The participants in the experience were a random sample of second-year computer science students from Algeria's University of Oum El Bouaghi. They were given two exercises to complete, the first of which asked them to write a C++ code that sorts three integers according to a given key sequence ($n = 33$). The goal of this practice is to help the student practice and perfect his usage of conditional construction.

In the second ($n = 35$) activity, participants are asked with building code to search in an array for a certain value. The goal of this activity is to provide the student with practice with data structures.

After gathering the relevant information, we had the twelve educators' rate and amend the 68 printed codes we had acquired. Second, we sent the identical set of 68 printed codes into both versions of the developed system and let them do the scoring automatically. Finally, we compared the instructors' ratings with the automated ratings we acquired using the developed systems.

5.2 Results

To prove which of the two developed versions of the system is more effective than the other, we used the statistical coefficient Spearman's rho. This statistical coefficient can be used to see how well the two versions of the developed system rank the students compared to teachers, by putting the following hypothesis:

H0: Is the developed system more likely to provide scores that are close to instructors' scores while using the hybrid matching algorithm or the ontologies-agent algorithm?

The results shown in Table 1 show that there is a high alignment between the two versions of the developed system and the average overall scores of the trainers, With the exception of three teachers in the first code (Expert 7, Expert 10, Expert 12) and one teacher in the second code (Expert 10), the Spearman's rho correlation was found to be not statistically significant.

After examining the results shown in Table 2, it is evident that there is a significant decrease in the coefficients of Spearman's rho values across all variables in the improved system compared to the corresponding values observed in the first version. This decrease is due to the fact that the improved system can identify two distinct categories of errors, namely syntactic and semantic errors, within students' codes. In contrast, the first version of the developed system can only detect syntactic errors, which have a direct impact on the overall scores. It can be concluded that the improved system shows greater accuracy compared to the first version, as shown by the results presented in Table 2.

The Pearson correlation coefficient is calculated for both scenarios involving the two codes. Our analysis of the findings indicates a large association between the developed system scores and instructors' overall average scores in both codes. Specifically, for the first code, a correlation coefficient of $rs = 0.67$ ($p < 0.05$, Table 1) and $rs = 0.64$ (p

Table 1. Correlations of teachers and the first version of the developed system

Teacher	Code 1	Code2
Expert 1	0.63***	0.43**
Expert 2	0.51**	0.42**
Expert 3	0.52**	0.40**
Expert 4	0.62***	0.49**
Expert 5	0.61***	0.56***
Expert 6	0.62***	0.39*
Expert 7	0.24 n.s	0.48**
Expert 8	0.53**	0.27 n.s
Expert 9	0.70***	0.36*
Expert 10	0.29 n.s	0.41*
Expert 11	0.69***	0.55**
Expert 12	0.27n.s	0.52**
Average	0.67***	0.62***

Table 2. Correlations of teachers and the improved system

Teacher	Code 1	Code 2
Expert 1	0.60***	0.42**
Expert 2	0.49**	0.41**
Expert 3	0.50**	0.39*
Expert 4	0.61***	0.47**
Expert 5	0.59***	0.55***
Expert 6	0.60***	0.37*
Expert 7	0.21 n.s	0.47**
Expert 8	0.52**	0.25 n.s
Expert 9	0.69***	0.35*
Expert 10	0.27 n.s	0.40*
Expert 11	0.67***	0.52**
Expert 12	0.24n.s	0.50**
Average	0.64***	0.59***

< 0.05, Table 2) was observed. Similarly, for the second code, a correlation coefficient of $rs = 0.62$ ($p < 0.05$, Table 1) and $rs = 0.59$ ($p < 0.05$, Table 2) was identified. It is

Table 3. Independent t-test between the improved system and teachers.

	Exercise 1		Exercise 2	
	2nd S_Onto_ALPPWA's	Teachers	2nd S_Onto_ALPPWA's	Teachers
Mean	10,025	09,98	09.96	09.94
SD(standard deviation)	2,1298	2,1725	2.1025	2.1000
T-test	0,1197			
Df	58			
p-value	0,9051			

SD: standard deviation, t-test: student test value, df: degree of freedom, P-value: probability value.

important to note that the distribution of the developed system and the average scores of the instructors align with the notion of normalcy.

It is important to note that we have already shown that the improved system exhibits a higher degree of precision compared to its first one. In the following findings, we only compare the improved system scores with the instructors' average scores.

We calculate the descriptive statistics between the improved system and teachers' overall average scores by using the statistic program SPSS and we compare the improved system and the average teachers' scores by student t-test.

We note from the results that the (P-value = 0,9051 > 0.05) then we can confirm that for the first code, there was no significant difference in the improved system's marks (Mean = 10,025, SD = 09,98) and the teachers' overall average scores (Mean = 09,98, SD = 2,1725, P-value = 0,9051 > 0.05). Also, in the second code, the same note for the (P-value = 0,9051 > 0.05). Consequently, there was no significant difference in the improved system's scores (Mean = 09.96, SD = 2.1025, P-value = 0,9051 > 0.05) and the teachers' overall average scores (Mean = .09.94, SD = 2.1000, p-value = 0.9051 > 0.05).

Figure 1 illustrates how the improved system's scores were positioned on the teachers' scoring range of each submission. So almost all the improved system's scores were in the scoring range Except for one case that was identified in the second code (submission number 6 in Fig. 1. Code 2).

The Hierarchical Ascendant Classification (HAC) technique was used in this study, using the SPSS statistical software. The clustering findings reveal that three separate groupings display different approaches toward scoring. By using distribution scores, we have identified the distinguishing features of three distinct scoring behaviors: lenient (CLU3_1 = 3), severe (CLU3_1 = 1), and moderate (CLU3_1 = 2). This information is visually shown in Fig. 2 and further consolidated in Table 4.

Three behaviors of instructors were recognized. The first set of instructors, as shown by the data presented in (Fig. 2) and Table 4 (CLU3_1 = 1), namely T1, T4, and T11, assigned lower scores in contrast to the lenient teachers (CLU3_1 = 3), namely T5, T6, and T10, who assigned higher values compared to the other teachers. The last cluster of instructors in the medium range (CLU3_1 = 2, as seen in (Fig. 2 and Table 4) (T2, T3,

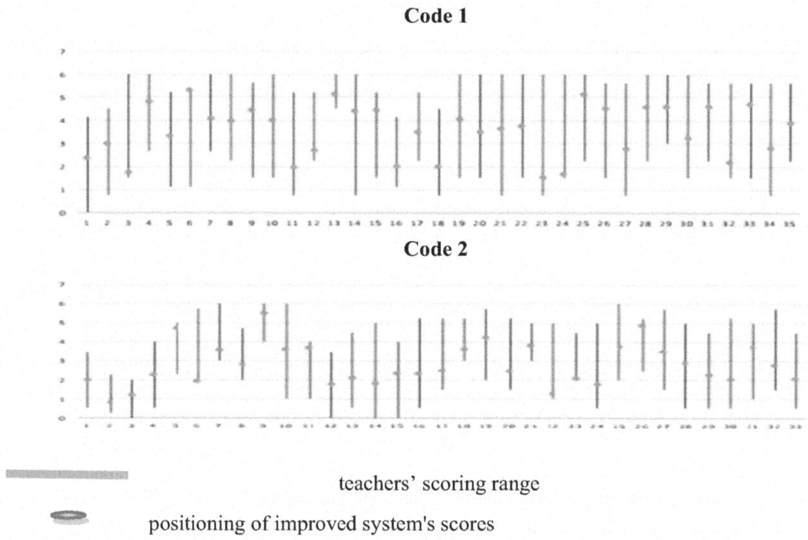

Fig. 1. The improved system's scores positioning on the teachers' scoring range of each submission.

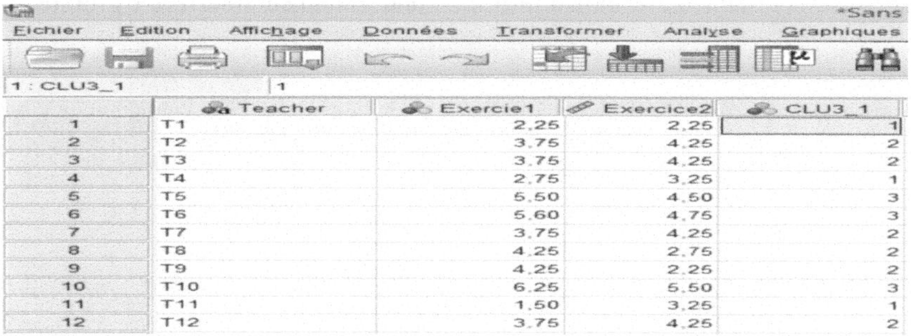

	Teacher	Exercie1	Exercice2	CLU3_1
1	T1	2.25	2.25	1
2	T2	3.75	4.25	2
3	T3	3.75	4.25	2
4	T4	2.75	3.25	1
5	T5	5.50	4.50	3
6	T6	5.60	4.75	3
7	T7	3.75	4.25	2
8	T8	4.25	2.75	2
9	T9	4.25	2.25	2
10	T10	6.25	5.50	3
11	T11	1.50	3.25	1
12	T12	3.75	4.25	2

Fig. 2. The behavior classification of the teachers in three classes (lenient, severe, middle)

Table 4. The behavior classification of the teachers in three classes (lenient, severe,middle)

Teacher	Teacher's class for the two exercises 1 and 2
T1, T4, T11	***Severe***
T2, T3, T7, T8, T9, T12	***Middle***
T5, T6, T10	***Lenient***

T7, T8, T9, T12) assigned scores that seem to be moderate, neither exceptionally high nor notably low (Fig. 3) illustrates the clustering of instructors in codes one and two.

We can notice from this teacher clustering (Fig. 3) that there are some changes in teacher behavior, for example, teacher (T4) changed from *severe* in the first code to the

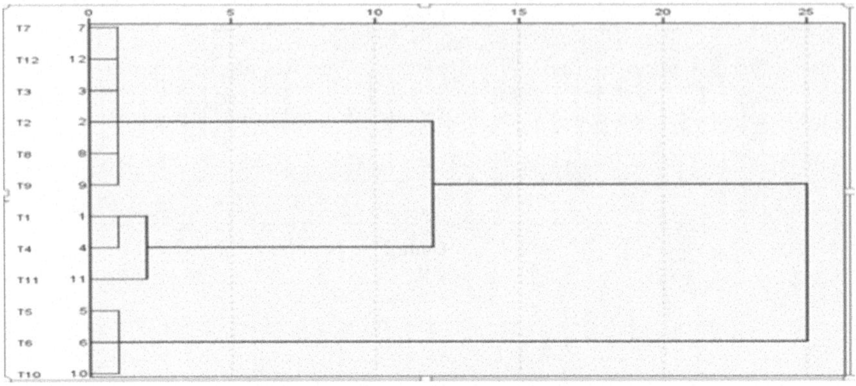

Fig. 3. Teachers' clustering in codes one and two

middle in the second code, and the same case for teachers (T3, T8, T9) their behavior changed from the *middle* in the first code to *severe* in the second code (exercise).

To confirm that the improved system's scores are given the same division of teachers' classes, we calculate the clustering suggested using the Hierarchical Ascendant Classification (HAC) with the SPSS statistic program, we use the same submissions of the teacher's results obtained are shown in (Fig. 4 and 5) and summarized by Table 5.

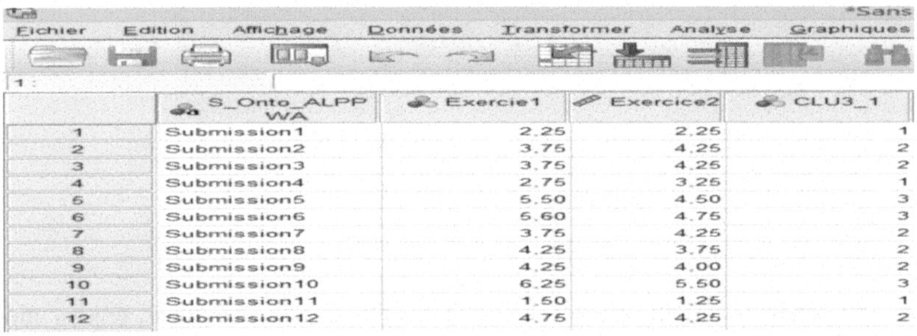

	S_Onto_ALPPWA	Exercie1	Exercice2	CLU3_1
1	Submission1	2.25	2.25	1
2	Submission2	3.75	4.25	2
3	Submission3	3.75	4.25	2
4	Submission4	2.75	3.25	1
5	Submission5	5.50	4.50	3
6	Submission6	5.60	4.75	3
7	Submission7	3.75	4.25	2
8	Submission8	4.25	3.75	2
9	Submission9	4.25	4.00	2
10	Submission10	6.25	5.50	3
11	Submission11	1.50	1.25	1
12	Submission12	4.75	4.25	2

Fig. 4. The behavior classification of the developed system in three classes(lenient, severe, middle)

Table 5. The behavior classification of the developed system in three classes(lenient, severe, middle)

Submissions number	S_Onto_ALPPWA class for the two codes 1 and 2
1, 4, 11	***Severe***
2, 3, 7, 8, 9, 12	***Middle***
5, 6, 10	***Lenient***

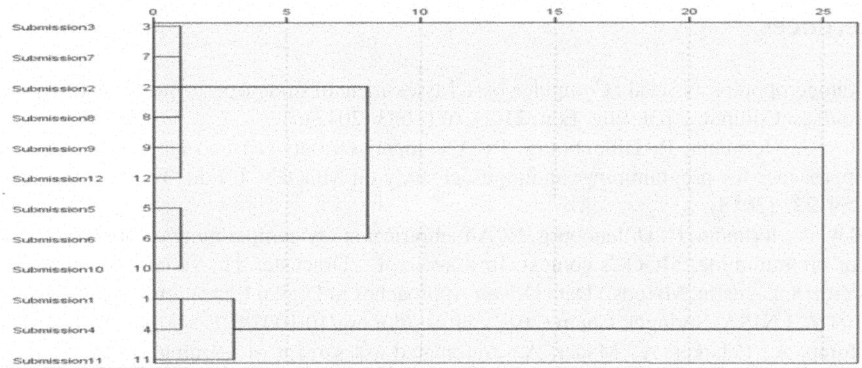

Fig. 5. S_Onto_ALPPWA clustering in codes one and two

We did not notice any differences in the improved system scoring that may distinguish it from teachers. So, the improved system kept the same behavior results, because it had a normal scoring behavior compared with teachers. This makes us make sure that the improved system is effective as a result of its ability and its suitability to the fully online assessment environment, and can it be completely relied upon in epidemic conditions for fully online assessment.

6 Discussion

The experimental results confirm that the improved system had identical scoring behavior to that of the instructors. The improved system demonstrates promise and ambition as a tool for facilitating the evaluation of students' computer programming codes. In comparison to traditional methods of student assessment, this tool offers improved service by alleviating teachers from the drawbacks associated with traditional assessment methods. Specifically, it reduces time requirements, minimizes effort, and enables accurate and objective evaluation. It is posited that the efficacy of the improved system has the potential to assist educators in accurately evaluating large classes.

7 Conclusion

This article demonstrates the study of how programming self-assessment emphasized student learning skills and faculty teaching skills compared to the traditional method. This article presents a case study that describes the transformation of the evaluation method of a programming course in higher education from a classic method to a fully online format during the COVID-19 pandemic, using a tool of the student-centered automated assessment called S_Onto_ALPPWA [5]. The study presented in this research has unveiled a novel direction for investigation, offering fresh recommendations to the continuing efforts aimed at enhancing future service provisions. To assess the potential impact of the second version of the developed system on student performance in programming, it is essential to conduct a comprehensive evaluation within the framework of Massive Open Online Courses (MOOCs).

References

1. Kalogeropoulos, N., et al.: Computer-based assessment of student performance in programing courses. Comput. Appl. Eng. Edu. **21**(4), 671–683 (2013)
2. Bey, A., Jermann, P., Dillenbourg, P.: A comparison between two automatic assessment approaches for programming: an empirical study on MOOCs. J. Edu. Technol. Soc. **21**(2), 259–272 (2018)
3. Bey, A., Jermann, P., Dillenbourg, P.: An empirical study comparing two automatic graders for programming. MOOCs context. In: Lavoué, É., Drachsler, H., Verbert, K., Broisin, J., Pérez-Sanagustín, M. (eds.) Data Driven Approaches in Digital Education, pp. 537–540. vol. 10474, LNISA, Springer, Cham (2017). https://doi.org/10.1007/978-3-319-66610-5_57
4. Rump, A., Fehnker, A., Mader, A.: Automated assessment of learning objectives in programming assignments. In: Cristea, A.I., Troussas, C. (eds.) Intelligent Tutoring Systems: 17th International Conference, ITS 2021, Virtual Event, June 7–11, 2021, 12677, LNPSE, pp. 299–309. Springer, Cham (2021). https://doi.org/10.1007/978-3-030-80421-3_33
5. Boussaha, K., Mokhati, F., Hanneche, A.: System-based ontology for assessing learner's programming practical works activities (S_Onto_ALPPWA). Int. J. Web-Based Learning Teach. Technol. **16**(5), 80–107 (2021). https://doi.org/10.4018/IJWLTT.20210901.oa5
6. Boussaha, K., Mokhati, F., Zakaria, C.: Architecture of a specific platform for training practical works: integration of learners' assessment component. Int. J. Technol. Enhan. Learn. **7**(3), 195–220 (2015)
7. Souza, D.M., Felizardo, K.R., and Barbosa, E. F.: A systematic literature review of assessment tools for programming assignments. In: 2016 IEEE 29th International Conference on Software Engineering Education and Training (CSEET). IEEE, pp. 147–156 (2016)
8. Keuning, H., Jeuring, J., and Heeren, B.: Towards a systematic review of automated feedback generation for programming exercises. In: Proceedings of the 2016 ACM Conference on Innovation and Technology in Computer Science Education, pp. 41–46 (2016)
9. Ullah, Z., et al.: The effect of automatic assessment on novice programming: strengths and limitations of existing systems. Comput. Appl. Eng. Edu. **26**(6), 2328–2341 (2018)
10. Lajis, A., et al.: A review of techniques in automatic programming assessment for practical skill test. J. Telecommun. Elect. Comput. Eng. (JTEC) **10**(2–5), 109–113 (2018)
11. Ismail, M.H., Lakulu, M.M.: A critical review on recent proposed automated programming assessment tool. Turkish J. Comput. Math. Edu. (TURCOMAT) **12**(3), 884–894 (2021)
12. Gupta, S.A.: E-assessment tools for programming languages: a review. In: Proceedings of the First International Conference on Information Technology and Knowledge Management, New Delhi, India, pp. 65–70 (2017)
13. Douce, C., Livingstone, D., Orwell, J.: Automatic test-based assessment of programming: a review. J. Edu. Resour. Comput. (JERIC) **5**(3), 4 (2005)
14. Combéfis, S.: Automated code assessment for education: review, classification and perspectives on techniques and tools. Software **1**(1), 3–30 (2022)
15. Author et al.: A novel approach based ontology for assessing learner's skills in the programming language practical works activities. In: the 3rd İnternational Conference on Advanced Reseach in Education Educatıonconf 2021 Oxford, UK (2021)

YOLOv3 for Brain Tumor Detection

Aida Chefrour[1,2](✉), Samia Drissi[1,3], Narimene Kechaichia[1], and Hadjer Haou[1]

[1] Computer Science Department, Mohamed Cherif Messaadia University, Souk Ahras, Algeria
{aida.chefrour,s.drici}@univ-soukahras.dz

[2] LISCO Laboratory, Badji Mokhtar University, Annaba, Algeria

[3] LiM Laboratory, Faculty of Science and Technology, University of Souk Ahras, P.O. Box 1553, 41000 Souk-Ahras, Algeria

Abstract. Brain tumors are fatal diseases generally caused by abnormal cells growing in the brain. Therefore, early and accurate detection of this disease can save the patient's life. This paper proposes a novel framework for the detection of brain tumors. The framework is based on the use of the YOLOv3 technique for detection. The proposed framework employs a deep neural network model called YOLOv3 for brain tumor detection. The Yolov3 technique is employed, using the principle of encoding bounding boxes to detect and predict the regions and types of brain tumors. Experiments with detection are performed on the Kaggle dataset of brain tumors. The results of experiments demonstrate that the proposed framework has achieved an 80% confidence score and an accuracy of 93.4% for the detection of brain tumors. The results demonstrate the efficiency of the proposed framework and have a higher performance level than other recent literature studies. Furthermore, this study is going to help in the diagnosis of brain tumors by doctors.

Keywords: Deep learning · YoLo · brain tumor

1 Introduction

Controlling the neural system, the eighth and most crucial organ in the human body is the brain. Brain tumors can result from the irregular and uncontrolled development of brain cells. The two basic types of brain tumors are primary and secondary tumors, which have the highest cancer death rates in the world, and neither their origin nor their growth rate is known. Primary tumors, such as brain tumors, form in the brain tissues; secondary tumors, on the other hand, form in another part of the body and travel by blood flow to the brain. Meningioma, glioma, and pituitary tumors are among the most harmful primary brain tumors, and they are the most challenging to detect and effectively treat. In addition, if treated, these may progress into critical situations [1].

The most important phase in diagnosis and medications to protect the patient's life is the early detection of brain tumors with high prediction accuracy. However, radiologists and other medical professionals must manually analyze brain MR images to find the tumor and healthy tissues, detect the tumor, and localize it. To resolve this problem, a system for computer-aided diagnostics is required. It must be utilized to reduce the effort

M. R. Laouar et al. (Eds.): ICISAT 2023, LNNS 981, pp. 61–75, 2024.
https://doi.org/10.1007/978-3-031-60591-8_6

and aid radiologists and other medical experts in medical images analysis. The task of automating the detection of brain tumors has been the focus of various research studies in the past [2].

Radiologists are using artificial intelligence in a variety of brain tumor research studies [3]. Deep learning techniques, such as convolutional neural network (CNN) architecture, are used for the identification of brain tumors in many different ways [4, 5].

Deep learning (DL) combines feature extraction and classification in a self-learning method, significantly improving performance. Additionally, CNN has made it evident that DL approaches perform better than traditional techniques to diagnose brain tumors from MR images. In the related work, DL performed the detection, and several pre-trained CNN models were employed for brain tumor detection. When tumors differ significantly in size, form, and intensity, identifying cancers can be difficult. The insufficient anticipation of this type of problem is due to the limitations of previous research.

Traditional techniques, which are aggressive and momentary, focus on the manual detection of tumor regions. Traditional algorithms and variations on them have failed to significantly improve performance. Furthermore, all tumor classifications were insufficiently represented in the limited, local datasets on which the prior approaches had been tested and trained.

In this paper, we propose a framework for brain tumor detection based on deep learning. We used the single-stage detector YOLO (You Only Look Once), which has a quasi-convolutive structure where a fixed number of bounding boxes will be provided for a given image. Yolo has demonstrated its performance in the detection of objects in real-time. The detection results obtained are very encouraging.

The following is a list of the major contributions to the work:

- For the study of brain tumors using MR images, a two-phase automated deep learning-based detection approach has been proposed.
- Implemented a YOLOv3 method based on a darknet-53 deep neural network that is capable of automatically locating tumors and their generated bounding boxes in images.

The remainder of the paper is structured as follows: In Sect. 2, related work is discussed. In the third section, we briefly survey the literature on the algorithms used. In Sect. 4, we describe in detail our contribution to detection. Sections 5, 6 describe the experiment we conducted and the results that our framework achieved. It also compares them with top-ranked algorithms. Finally, we draw some conclusions and show ongoing research aspects in Sect. 7.

2 Related Works

In recent years, several studies have been done for the detection of brain tumors using deep learning. We have chosen to discuss in this section five very recent works described in the literature:

- To identify brain tumors (BT) using MRI scan datasets, researchers from [6] developed a transfer learning (TL) approach for deep learning models with fine-tuning. They proposed a YOLO-based model for MRI tumor detection. A dataset of 3064 T1-weighted, contrast-enhanced (CE) MRI scans that were preprocessed and tagged using a tool was used to train the YOLO (version-4) model for the technique. With the aid of TL, we had the benefit of pre-trained dataset loading from COCO, which accelerated learning and provided the necessary features for identifying brain tumors. The displayed piece was created using a 29-layer YOLO. Tiny and precisely crafted to function effectively and complete tasks in the majority of circumstances with good execution. The model's results are better to those of prior iterations of YOLO and other studies like Fast R-CNN in terms of precision, recall, and F1-score.
- [7] Suggested transfer learning methods for a deep learning model to use MRI data to find malignant tumors, such as glioblastoma. This study uses the advanced object detection framework YOLO (You Only Look Once) to provide a deep learning-based method for identifying brain tumors. YOLOv5 is a modern deep learning method for object recognition that requires less computing infrastructure than other similar models. This study made use of the Brats 2021 dataset from the RSNA-MICCAI brain tumor radiogenetic classification. The MakeSense machine learning network tool for labeling datasets has been used to label images from the RSNA-MICCAI brain tumor radiogenetic competition dataset. The preprocessed data is then divided into the model's training and testing sets. The YOLOv5 model has an accuracy rate of 88%. Their model is tested using the entire data set, and it is discovered to be successful in identifying brain tumors.
- This work [8] showed how a portable microwave head imaging device (MWHI) and the deep learning-based YOLOv5 object detection system can automatically diagnose human brain diseases. 400 RMW image samples, including non-tumor and tumor samples, were initially collected using the developed MWHI system in diverse locations. The RMW image has a resolution of 640×640 pixels. Then, using image pre-processing and augmentation methods, the training dataset, consisting of 4400 images, is created. Models are later tested on 20% of the images after being trained on 80% of them. In the future, the models will be validated using 20% of the training dataset. The detection performance is assessed using the YOLOv5 model iterations YOLOv5s, YOLOv5m, and YOLOv5l. The study found that the YOLOv5l model performed better than the YOLOv5s and YOLOv5m variants. In terms of accuracy, precision, sensitivity, specificity, F1-score, mean average precision (mAP), and classification loss, the YOLOv5l model achieves 96.32%, 95.17%, 95.28%, 95.53%, 96.12%, and 0.0130, respectively. The YOLOv5l model automatically detected tumors in RMW images and distinguished between benign and malignant tumors using a predicted bounding box and objectness score. The YOLOv5l object detection model can thus be recognized for automatic tumor diagnosis and classification in a portable microwave brain imaging device as a real-time application.
- A machine learning system entitled YOLO v5 SSD (single shot detection) was proposed in this work [9] to detect and categorize cancers like meningiomas, gliomas, and pituitary glands with 88% accuracy. The open-access Kaggle dataset was enhanced with data for this purpose. A total of 399 images of pituitary tumors and 396 images each of gliomas, meningiomas, no tumors, and other MRI classifications were used.

The performance of the You Only Look Once (YOLO) v5 classifier was evaluated using the data from the current study's false negative, true positive, false positive, and true negative categories. It was demonstrated that the Yolo v5 model offered 88% accuracy.

- The authors of [10] developed a technique for the diagnosis and localisation of brain tumors by proposing an automated methodology for these tasks. Magnetic resonance imaging analysis is used in the approach. By exploiting the information provided by medical images of the brain, the proposed technique seeks to enhance the diagnosis and precise localisation of brain cancer in order to improve patient prognosis and treatment outcomes. They automatically identify and locate brain tumors using the YOLO approach. In the study of 300 brain pictures, we found that the proposed model for brain cancer detection and localization was effective with a precision of 0.943 and recall of 0.923 for brain cancer detection and a mAP_0.5 equal to 0.941 for brain cancer localization and detection.
- A paradigm to identify and classifying cancers that can be evident on brain MRI imaging is presented in this work [11]. The YOLO-v3 object detection technique is utilized for detection. The target topic is swiftly determined in its third iteration, YOLOv3 (you look only once), and a confidence rating is provided. Using a convolutional neural network (CNN), the detected tumor is then categorized into low-grade and high-grade gliomas in accordance with the WHO grading system. The effectiveness of the suggested strategy is assessed, and the classification precision and results are contrasted with those of other state-of-the-art models. The proposed method employing CNN and YOLOv3 offers a higher probability of correctly recognizing the brain tumor with an overall accuracy of 97%.
- The authors of [12] used the YOLOv3 approach to detect brain tumors in cancer patients. The YOLOv3 model was trained using a set of 3064 T1-weighted contrast-enhanced (CE) MRI images that had been previously processed and labeled. The MSCOCO dataset's pre-trained weights were combined using transfer learning to produce a potent set of features for brain tumor detection. With a mean average precision of 94.14%, a precision of 91.34%, a recall of 90.58%, and an F1-Score of 92.55%, they outperformed earlier YOLO detection algorithms and research that employed bounding box detection to complete the same objective. This is a significant achievement.

3 Background

Our proposed framework comprises Yolov3, which discovers a localization for detection. In this section, we briefly review the main concept of Yolo:

"You Only Look Once" (YOLO) is a common algorithm that is used in numerous object detection applications. Yolo is well-known for its speed and accuracy in real-time object detection. Yolo uses convolutional neural networks (CNNs) to effectively find objects. Yolo needs to complete the first phase of CNN in order to classify objects and identify them using colored boxes. This method divides the image into regions and predicts the probability for each one, as seen in Fig. 1. This technique occurs in a variety of iterations, including Yolo V2, Yolo V3, Yolo V4, and Yolo V5 [13]. The Yolo architecture consists of 24 convolutional layers and two fully connected layers. Figure 2 shows the basic architecture.

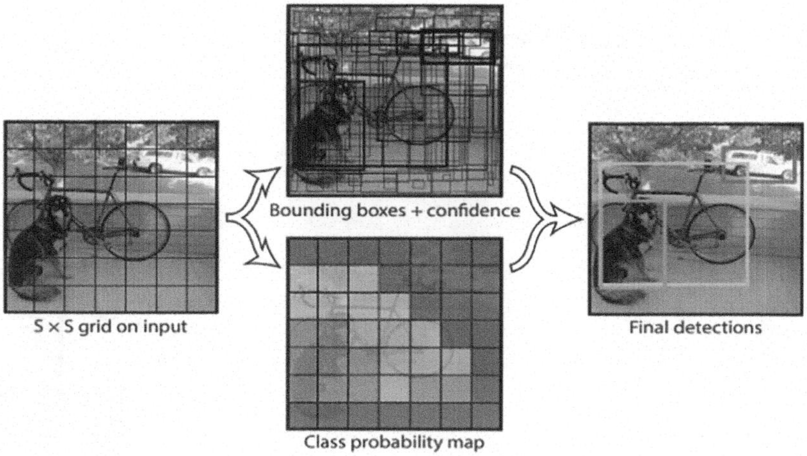

Fig. 1. Object Detection and Classification using YOLO [14].

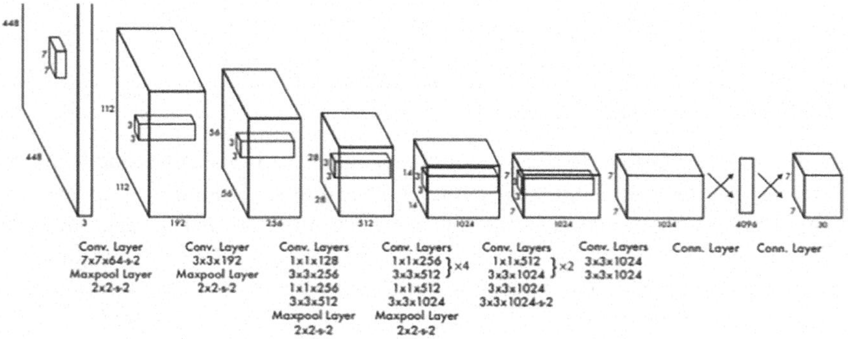

Fig. 2. YOLO architecture [14].

4 Proposed Algorithm

In this section, we discuss the proposed framework's deep architecture. For the purpose of detecting brain tumors, the suggested framework uses a deep neural network model (YOLOv3). Figure 3 shows the proposed structure to detect brain tumors.

Joseph Redmon et al. developed YOLO (You Only Look Once), one of the most effective real-time object detection systems [15]. Currently, a super-fast, sophisticated object detection method is available called YOLO version 3 (YOLOv3) [16]. It builds on the ideas of YOLO [15] and YOLOv2 [17]. When the input image is partitioned into a grid of cells, each cell in YOLOv3's single DCNN directly predicts a boundary box (BB) and object classification.

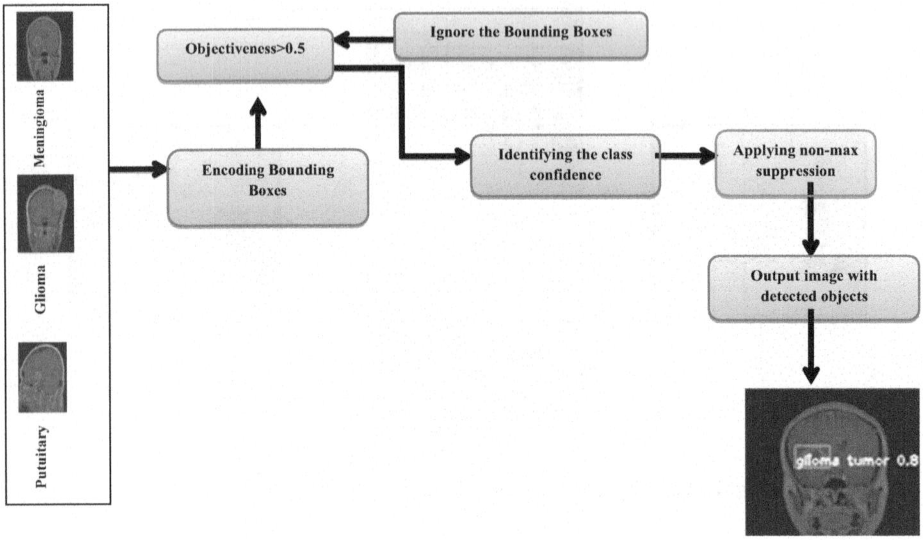

Fig. 3. Proposed detection technique using YoLo.

4.1 Training Details

We used Yolov3 technology to train a model that can identify the types and locations of brain tumors. Yolo needs an input consisting of the testing and training components from an image data set. The location of the tumor frame in the original image, the image's width and height, and a private key that identifies the type of this brain tumor must also be attached to each image in a file with the same name as the photo. We receive a trainer model that is ready to use after the training procedure. In the prediction phase, we just input one image into the model, and the results take the shape of a colored box that identifies the tumor's location in the image and a confidence score that indicates the general type of predicted brain tumor. The main proposed architecture is shown in Fig. 3.

For More Explanation

The algorithm's basic concept is to create a M*M grid from the input image and make predictions in each cell of the grid. Additionally, if the location of an object's center is identified in a grid cell, then that grid is in charge of identifying that object. B bounding boxes are projected in each grid cell, together with their confidence scores. These confidence scores also indicate whether an object is present in the grid cell and how well the box predicted it. The confidence score is typically calculated as follows:

$$\text{Confidence D Pr (Object) IoU (Grount Truth; PredBox)}$$

The intersection over the union of the GroundTruth (GT) and predicted box (PredBox) is represented by Pr (Object) Î (0,1), IoU. If an item does not exist in that cell, the confidence score (C) is equal to the intersection over union ("IoU") between the GT and PredBox; otherwise, it should be "1". Each cell also forecasts the probabilities of the C conditional class for the identified object. Therefore, the conditional class probabilities bx, by, bw, bh, and pc are the total (5 + C) values expected by each cell.

In this instance, bx represents the box's center position coordinates, and bw and bh represent the box's width and height.

Additionally, by accurately sampling the input image sizes by 32, 16, and 8, YOLOv3 predicts objects at three different scales. The problem of object size changes might be resolved in this way. But the prediction from YOLOv3 is encoded as:

$$M \times M \times (B \times 5 + C)$$

The number of bounding boxes in a cell depends on the image that was entered into the grid cell, with 5 representing 4 bounding box characteristics, 1 object confidence, and the number of classes. Multiple classes can apply to a target object. Consequently, a convolutional layer and a logistic regression activation function are utilized as part of the SoftMax layer. A threshold detects the target object, and logistic regression projects the confidence score for each class. To determine the difference between the projected value and the actual value, three loss functions—classification loss, localization loss, and confidence loss—are typically applied.

4.2 Encoding Bounding Box

To fully understand this part of the YOLO algorithm, it is crucial to describe what is now expected. The fact that it just uses one convolutional network puts it above most other neural network models. The bounding boxes and resulting probability are predicted by the convolutional network. The probabilities are used to weight the bounding boxes. The model makes the final weights a determining factor in their detection. This makes it possible to directly maximize the performance of the model as a whole. Consequently, images may be created and processed quickly [18]. Each bounding box can be described by one of the following four terms:

1. value 'c' refers to an object class;
2. center of bounding box (bx, by)
3. width (bw)
4. height (bh)
5. It is also necessary to predict the value of pc, which indicates the probability that there is an object inside the bounding box [18].

As shown in Fig. 4, YOLO separates an input image into grids (such as a 3 by 3 grid) after taking an input image.

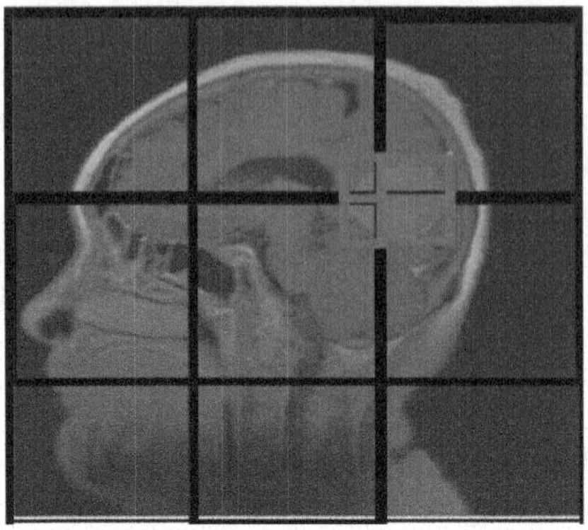

Fig. 4. Input image divided into 3 * 3 grid.

On each grid, image localization is used. Then, YOLO predicts the bounding boxes for objects and the class probabilities that go along with them. The labeled data must be provided to the model during training. In our example, there are a total of three classes into which the items must be classified, and the image is divided into a 3 × 3 grid. Meningioma, glioma, and putuitary are our three classes. According to Table 1, the y-label is a five-dimensional vector for each grid cell.

Table 1. Dimensional vector for each grid cell.

Y =	c = 0
	Bx = 0.689453, by = 0.381836
	Bw = 0.183594
	Bh = 0.197266
	Pc = 1

4.3 Making Predictions

The conversion of the network's output into bounding box predictions is described in the following formulas: The x, y, width, and height of our prediction are represented by the variables bx, by, bw, and bh. The output of the network is bx, by, bw, and bh. Cx and Cy are the top-left network coordinates. Anchor box dimensions are pw and ph [18]:

$$bx = \sigma(tx) + cx$$

$$by = \sigma(ty) + cy$$

$$bw = pw \times etw$$

$$bh = ph \times eth$$

Given that there is an object in this grid that is a glioma tumor, pc = 1 (in Fig. 4), and c1 = 1 as a result. The x and y coordinates of the grid object's center in YOLO are the grid coordinates of the form bx, by. In this instance, the values will be around bx = 0.689453 and by = 0.381836. The grid's center will always be in the grid, while bx and by will always be in the range of 0 and 1.

4.4 Intersection Over Union and Non-max Suppression

It quantifies the point where the expected and real bounding boxes intersect over the union. The IoU determines the area of the intersection over the union of the two boxes.

In general

$$\text{IoU} = \text{area of intersection/area of union}$$

Estimating is reasonable if IoU is higher than 0.5. A problem that explicitly states that it may cause the limit of 0.5 to be modified. It makes logical sense that better predictions will result from increasing the boundary. Non-max suppression is an additional strategy that may significantly improve YOLO's yield. The ability of object detection algorithms to identify an object more than once rather than just once is one of their most well-known problems.

Non-Max Suppression is used to get a single detection of any object, and the following approach is proposed:

1. First, the probability of each detection is considered and the highest one is selected.
2. The method then looks at the other boxes in the frame. Boxes that have a high IoU with the currently selected box are removed.s
3. After the boxes have been removed, a box is selected from all the most possible boxes, which has a high probability of removed boxes and less of the selected highest probability.
4. It again looks at the IoU of the box and compresses boxes with a high IoU.

Steps 1 through 4 of the procedure are repeated until the final bounding boxes are chosen or compressed. After performing the previous steps, only object detection is obtained. Non-max suppression is a method of selecting the boxes and suppressing the nearby boxes with non-max possibilities with the highest probability.

4.5 Anchor Boxes

The idea of anchor boxes is introduced when there are multiple objects in a single grid. An object is allocated to a grid by taking its approximate position and is then assigned to the corresponding grid based on its position in the 3 × 3 grid. There is difficulty if the centers of the two objects occur within the same grid. For the construction of both boxes, there are two predetermined methods known as anchor boxes or anchor box shapes. Instead of obtaining only one output for each grid, we obtain two.

5 Experimental Setup

5.1 Dataset

We collected a brain tumor data set of normal and tumor images from the open-source Kaggle website [19]. It contains 3064 images of 233 patients with a dimension of 512 × 512 and three tumor types: (1) meningioma, (2) glioma, and (3) pituitary. This database consists of two sets of images in PNG format: a training set of 2450 images and a test set of 614 images. Remember that an image is a collection of pixels. Each pixel is a triplet of values (R, G, and B) (see Fig. 5).

(A)

(B)

(C)

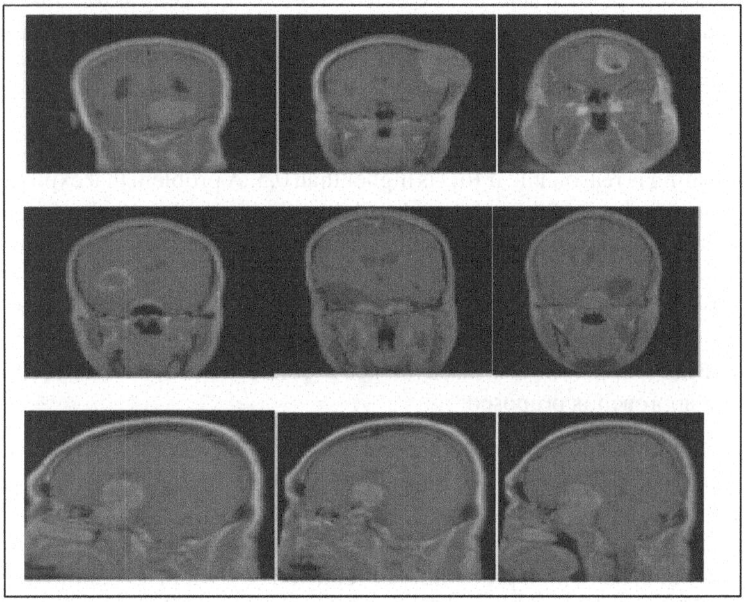

Fig. 5. (A) Meningioma, (B) glioma, and (C) pituitary tumor images are examples from the data set.

5.2 Performance Evaluation Parameters

By performing significant experimental simulations, we verify the proposed approach's effectiveness and robustness. Accuracy (Acc), precision, recall, and F1-score are four commonly used evaluation metrics on which quality is measured. The equations in Table 2 and the following list can be used to calculate these parameters:

$$\text{Accuracy} = \frac{TP + TN}{TP + FP + TN + FN}$$

$$\text{Precision} = \frac{TP}{TP + FP}$$

$$\text{Recall} = \frac{TP}{TP + FN}$$

$$\text{F1\_score} = \frac{2 \times Precision \times Recall}{Precision + Recall}$$

Table 2. Description of different evaluation metrics

.Metrics	Description
Precision	Ratio of successfully identified classes to total classes:
Recall	The percentage of classes correctly identified and real negative classes
F1-score	The harmonic average of precision and recall
Accuracy	% of total correct detections
TP	Correctly positive prediction
TN	Correctly negative prediction
FP	Incorrectly positive prediction
FN	Incorrectly negative prediction

6 Results and Discussions

6.1 Data Preprocessing

The tumor location is identified in each image using a special program called LabelImg (Fig. 6), where we determine the position of the tumor in each image, and the application creates a file with a name similar to the image that includes the hand's bounding box's height and width as well as its placement inside the image (Fig. 7).

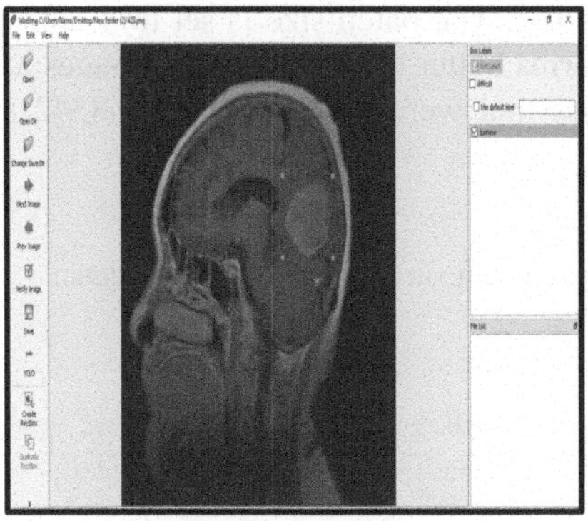

Fig. 6. The LabelImg interface.

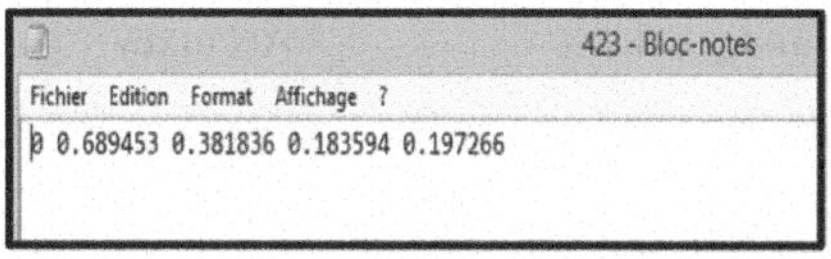

Fig. 7. File created through the labelImg application.

6.2 Classification Process

We use 80% of the training stage for the pre-processing of data and 20% for the test (as mentioned in the previous section).

To verify that all issues are present in both the training and test parts, we partition the data set. The number of layers and the number of filters in each class were modified from the original Yolo structure. We create a Yolo model during the training process using the training part of the data. Using Darknet53 weights (Figs. 8 and 9 show the full architecture) during the training process. When the value of the loss exceeds its maximum, the training will be brought to an end. The results obtained are: the bounding boxes detected at the last layer of the model that exceed the degree of confidence > 0.2 with the name of the class and the degree of confidence (see Fig. 10).

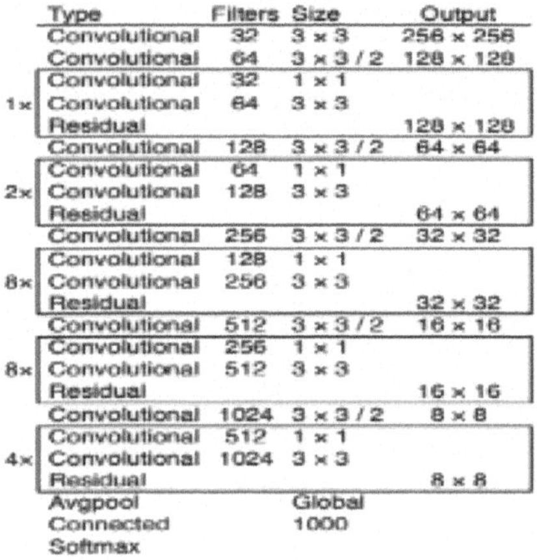

	Type	Filters	Size	Output
	Convolutional	32	3 × 3	256 × 256
	Convolutional	64	3 × 3 / 2	128 × 128
1×	Convolutional	32	1 × 1	
	Convolutional	64	3 × 3	
	Residual			128 × 128
	Convolutional	128	3 × 3 / 2	64 × 64
2×	Convolutional	64	1 × 1	
	Convolutional	128	3 × 3	
	Residual			64 × 64
	Convolutional	256	3 × 3 / 2	32 × 32
8×	Convolutional	128	1 × 1	
	Convolutional	256	3 × 3	
	Residual			32 × 32
	Convolutional	512	3 × 3 / 2	16 × 16
8×	Convolutional	256	1 × 1	
	Convolutional	512	3 × 3	
	Residual			16 × 16
	Convolutional	1024	3 × 3 / 2	8 × 8
4×	Convolutional	512	1 × 1	
	Convolutional	1024	3 × 3	
	Residual			8 × 8
	Avgpool		Global	
	Connected		1000	
	Softmax			

Fig. 8. DarkNet 53

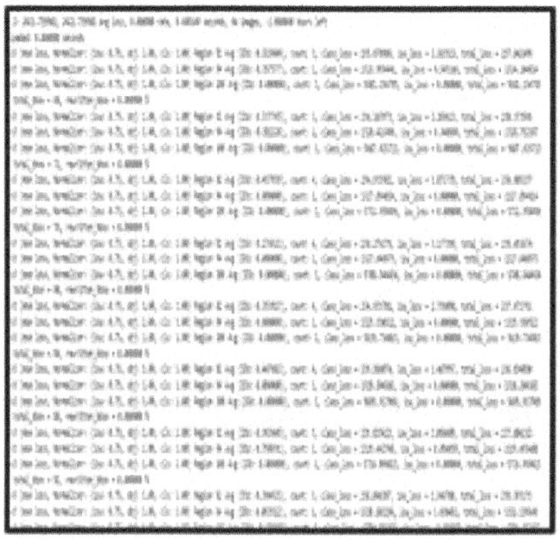

Fig. 9. Training process

Using an initial learning rate of 0.001, the model is trained for 200 epochs and then divided by four epochs. The learning rate is adjusted, which improves training accuracy and reduces validation loss.

The momentum optimizer has a setting of 0.2. The batch size is set to 4, and the weight decay is set at 0.0005. The model performs better for the provided images after 15000. As consequence, the accuracy and loss that were obtained were 0.8342 and 0.0934, respectively (see Table 3).

Table 3. The metrics of detection performance using various training sets of data.

Training dataset	Accuracy	Recall	F1-score
800	93.4%	91.85%	92.68%

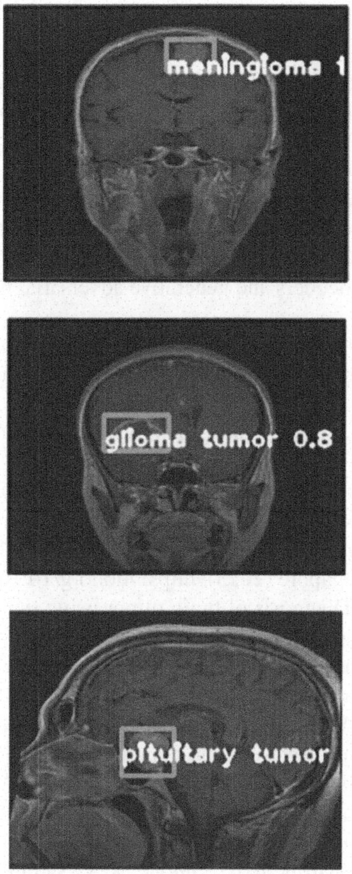

Fig. 10. Tumor detection results of the YOLOv3 model.

7 Conclusion and Perspectives

The patient's early treatment requires an effective technique for diagnosing brain tumors. To enhance brain tumor diagnosis and lower computing complexity, a new framework for brain tumor detection is being created. Using advanced real-time object identification algorithms (YOLOv3), we proposed a unique method for identifying and classifying different types of brain tumors based on score confidence and accuracy. Experimental results demonstrate that Yolov3 outperformed by achieving high accuracy (93.4%) and a score confidence of 0.8. The two-phase approach is planned to assist radiologists in the diagnosis of brain tumors as well as doctors in clinical practice decision-making.

In the future, we will use other versions of YOLO, such as YOLO v4, YOLO v5, and another database that contains four classes, for example, glioma, meningioma, pituitary, and non-tumor.

References

1. Abiwinanda, N., Hanif, M., Hesaputra, S.T., Handayani, A., Mengko, T.R.: World congress on medical physics and biomedical engineering. In: Brain tumor classification using convolutional neural network. IFMBE, vol. 68/1, pp. 183–189. Springer, Prague (2019). https://doi.org/10.1007/978-981-10-9035-6_33
2. Afshar, P., Mohammadi, A., Plataniotis, K.N.: Bayescap: a bayesian approach to brain tumor classification using capsule networks. IEEE Signal Process. Lett. **27**, 2024–2028 (2020)
3. Ahmad, B., Sun, J., You, Q., Palade, V., Mao, Z.: Brain tumor classification using a combination of variational autoencoders and generative adversarial networks. Biomedicines **10**(2), 22330 (2022)
4. Badža, M.M., Barjaktarović, M.Č: Classification of brain tumors from mri images using a convolutional neural network. Appl. Sci. **10**(6), 1999 (2020). https://doi.org/10.3390/app10061999
5. Bengio, Y., et al.: Learning deep architectures for AI. Found. Trends. Mach. Learning **2**(1), 1–127 (2009)
6. Dixit, A., Singh, P.: Brain tumor detection using fine-tuned YOLO model with transfer learning. In: Gupta, M., Ghatak, S., Gupta, A., Mukherjee, A.L. (eds.) Artificial Intelligence on Medical Data: Proceedings of International Symposium, ISCMM 2021. LNCVB, vol. 37, pp. 363–371. Springer, Singapore (2023). https://doi.org/10.1007/978-981-19-0151-5_30
7. Shelatkar, T., Bansal, U.: Diagnosis of brain tumor using light weight deep learning model with fine tuning approach. In: Singh, P., Singh, D., Tiwari, V., Misra, S. (eds.) Machine Learning and Computational Intelligence Techniques for Data Engineering. LNEE, vol. 998, pp. 105–114. Springer, Singapore (2023). https://doi.org/10.1007/978-981-99-0047-3_10
8. Hossain, A., Islam, M.T., Almutairi, A.F.: A deep learning model to classify and detect brain abnormalities in portable microwave based imaging system. Sci. Rep. **12**(1), 6319 (2022)
9. Raza, A.R., Amjad, U., Abubakr, M., Abbasi, A., Azam, H., Ali, A.: Multiclass light weight brain tumor classification and detection using machine learning model Yolo 5. UMT Artif. Intell. Rev. **2**(2), 25 (2022)
10. Mercaldo, F., Brunese, L., Martinelli, F., Santone, A., Cesarelli, M.: Object detection for brain cancer detection and localization. Appl. Sci. **13**(16), 9158 (2023)
11. Chanu, M.M., Singh, N.H., Muppala, C., Prabu, R.T., Singh, N.P., Thongam, K.: Computer-aided detection and classification of brain tumor using YOLOv3 and deep learning. Soft Comput. **27**(14), 9927–9940 (2023)
12. Kalyani, B.J.D., Meena, K., Murali, E., Jayakumar, L., Saravanan, D.: Analysis of MRI brain tumor images using deep learning techniques. Soft Comput. **27**, 7535 (2023)
13. Zhao, Z.Q., Zheng, P., Xu, S.T., Wu, X.: Object detection with deep learning: a review. IEEE Trans. Neural Netw. Learning Syst. **30**(11), 3212–3232 (2019)
14. Redmon, J., Divvala, S., Girshick, R., Farhadi, A.: You only look once: unified, real-time object detection. In: Proceedings of the IEEE Conference on Computer Vision and Pattern Recognition, pp. 779–788 (2016)
15. Redmon, J., Farhadi, A.: Yolov3: an incremental improvement. arXiv preprint arXiv:1804.02767 (2018)
16. Redmon, J., & Farhadi, A.: YOLO9000: better, faster, stronger. In: Proceedings of the IEEE Conference on Computer Vision and Pattern Recognition, pp. 7263–7271 (2017)

17. Kumar, C., Punitha, R.: Yolov3 and yolov4: multiple object detection for surveillance applications. In: 2020 Third International Conference on Smart Systems and İnventive Technology (ICSSIT), pp. 1316–1321, IEEE (2020)
18. Hassan, N.I., Tahir, N.M., Zaman, F.H.K., Hashim, H.: People detection system using YOLOv3 algorithm. In: 2020 10th IEEE International Conference on Control System, Computing and Engineering (ICCSCE), pp. 131–136, IEEE (2020)
19. https://www.kaggle.com/datasets/denizkavi1/brain-tumor

State of the Art on Terminological Resources from Field of HIV/AIDS for Monitoring People Living with HIV (PLHIV)

Camir Augustin Nacangny Malack(✉), Ibrahima Diop, Youssou Dieng, and Mouhamadou Gaye

University Assane Seck Ziguinchor, Ziguinchor, Senegal
c.malack2903@zig.univ.sn

Abstract. After the observation of the difficulties encountered in as part of the follow-up monitoring people living with HIV (PLHIV), an IT platform was set up for a concerted management and monitoring of PLHIV in the border area between Senegal, Gambia and Guinea-Bissau. The modules in development on the platform require a domain ontology of the HIV/AIDS to be put in place. This article not only describes the needs that motivate the creation of such an ontology, but also, makes the state of the art of termino-ontological resources (TOR) for HIV/AIDS and the identification of resources that can be reused in the construction of that ontology.

Keywords: HIV/AIDS · follow-up PLHIV · termino-ontological resources

1 Introduction

According to [1], the migration to Senegal over the years the 1970–80s, from Guinea, Guinea-Bissau, Gambia and the Senegal River valley (Mauritania) has had as its consequences, an intermingling of the population and economic and social activities shared through the borders, so making the monitoring of people living with HIV (PLHIV) in this area very difficult.

The dynamism and intensification of efforts to combat HIV/AIDS (Human Immunodeficiency Virus/Acquired Immunodeficiency Syndrome) for the reaching of the objectives the UNAIDS[1] at zone level border areas of West Africa are being reinforced by initiatives such as the FEVE[2] project (Frontiers and Vulnerability to HIV/AIDS in West

[1] **Objectives the UNAIDS**: that by 2025, 95% of people living with HIV know their HIV status, 95% of people who know their status are receiving treatment and 95% of people on HIV treatment have a suppressed viral load so their immune system remains strong and the likelihood of their infection being passed on is greatly reduced.

[2] **FEVE:** (Borders and vulnerability to HIV in West Africa) is developed by ENDA Santé to reduce the vulnerability, transmission and impact of HIV / AIDS among populations in high mobility areas and cross-border areas. The FEVE project area consists of 9 countries, which are: Cape Verde, Senegal, Mali, Niger, Burkina Faso, Guinea, Guinea-Bissau and Gambia, Ivory Coast.

M. R. Laouar et al. (Eds.): ICISAT 2023, LNNS 981, pp. 76–87, 2024.
https://doi.org/10.1007/978-3-031-60591-8_7

Africa), which focuses on the resolution of the difficulties encountered at monitoring people living with HIV in border areas. This project, through the activities of the NGO ENDA-santé[3], has enabled to the setting up of a social and semantic web platform [2] of cross-border[4] monitoring of PLHIV. This platform has dematerialised the HIV patient reached referral system in a two-page paper form [31] containing all the patient's clinical information. The platform settles then the security issues linked to the fact that the mobile patient between borders had to move with his form to enable the welcoming doctor to continue the follow-up in the best possible conditions. This platform is not only a sub-regional PLHIV monitoring system in three languages but also offers a communication and capacity building system between actors [3].

According to its architecture presented in [2], the platform monitoring of PLHIV filled shutters two.

- **social web**, which allows to accompany the exchange of patients between the doctors and to alert the later to referrals and self-referrals in the FEVE zone; this social web allows also to promotes discussions (by a private forum for doctors and a public forum for all actors) and to report lost to follow-up.
- **semantic web**, which makes it possible first to open up the data from this system for monitoring PLHIV in RDF format, second to link that data by knowledge of the domain of HIV and the process of monitoring PLHIV, third make more stable statistics on lost to follow-up and the semantic search thanks to a semantic search inference engine, and finally to predict lost to follow-up.

In order to implement the semantic part of the platform, it is necessary to dispose knowledge of the domain from HIV/AIDS. For this, the construction of this knowledge requires prior work which consists of identifying the terminological and ontological resources on HIV/AIDS. For this, the construction of this knowledge requires prior work which consists of identifying the terminological and ontological resources on HIV/AIDS. That constitutes the objective of our present article, because it allows us to know the existing on knowledge resources HIV/AIDS so that we may propose an HIV/AIDS domain ontology for monitoring PLHIV.

The next part of this article is organised as follows: Sect. 2 describes the needs for building an HIV/AIDS domain ontology, for monitoring PLHIV. The Sect. 3 presents a literary review on TOR that exist on the disease for HIV/AIDS. In Sect. 4, we do a discussion to show the importance of these existing termino-ontological resources (TOR) for the construction of the ontology. Finally, Sect. 5 concludes this article and is an announcement in prospect on the construction of the domain ontology, which will be the subject of another article.

[3] **NGO ENDA-santé**: https://www.enda-sante.org/fr/.

[4] **cross-border platform**: https://crossborder-platform.org/.

2 Description of the Need for an HIV/AIDS Ontology for Monitoring PLHIV

The notion of ontology in computing emerged around the 1990s. One of the most widely used definitions is that of Studer [4], who defines it as being « *An ontology is a formal, explicit specification of a shared conceptualisation*». In other words, an ontology allows to describe in a formal language the mental representation of objects defined in a consensual way in relation to a need.

The description of the needs which justify the creation of an HIV/AIDS domain ontology for monitoring PLHIV is given below:

- **Link PLHIV monitoring data** (i.e. linking references and self-references by the knowledge's of domain of HIV/AIDS): It is necessary to identify the concepts used in the referencing or self-referencing of a patient. This will enable to make more performing information retrieval, statistics, inference, prediction, etc. on the monitoring data of patients living with HIV/AIDS in the cross-border platform.
- **To Building a process ontology for monitoring PLHIV**: A process or task ontology allows to model the knowledge specific to the elaboration of a particular task by report to a domain. So, the PLHIV monitoring process ontology is based on the HIV/AIDS domain ontology and describes patient monitoring tasks. In so doing, we will be able to: list the possible treatments for a given case, validate or invalidate treatments.
- **To Describing and sharing resources monitoring PLHIV**: For resource descriptions, in addition to the description metadata, we use the HIV/AIDS domain ontology to make it easier to enter keywords. But it will also be used to link these keywords. For example, for a given resource in its description, we can have as the keywords: opportunistic disease, co-infection, tuberculosis, etc., which are terms equivalent to concept labels in the HIV/AIDS domain ontology.

The description of the needs of the HIV/AIDS domain ontology for monitoring PLHIV is based on the form reference [31] which was established as part of the FEVE project for monitoring PLHIV. This form, containing a patient's essential information, was the subject of a consensus between monitoring actors. The exploitation of the form has permitted to identify the concepts and their sub-domains as illustrated in Fig. 1.

As illustrated in Fig. 1, the sub-domain cutting dictated by the reference form has made it possible to restrict the HIV/AIDS domain ontology to the following sub-domains:

- **Biology**, it's the science that examines life in all its facets, from the microscopic cell to planetary ecosystems, to understand how organisms function and adapt to their environment [5]. By report, to our form reference we find there concepts and relationships relating to cd4, lymphocytes, haemoglobins, transaminases, etc.
- **Disease**, it's a generic term with blurring and shifting contours, that express a state of alteration in physical or mental health, or even social [6]. However, if we base at our form, we find there concepts and relationships relating to HIV/AIDS, co-infectious diseases, opportunistic diseases, etc.
- **Treatment**, it's a set of measures applied by a health professional to a person in the face of a disease, to help cure it, relieve its symptoms or prevent its onset [7]. So, our factsheet regroup concepts and relationships relating to antiretroviral, drugs and the treatment of co-infectious and opportunistic diseases, etc.

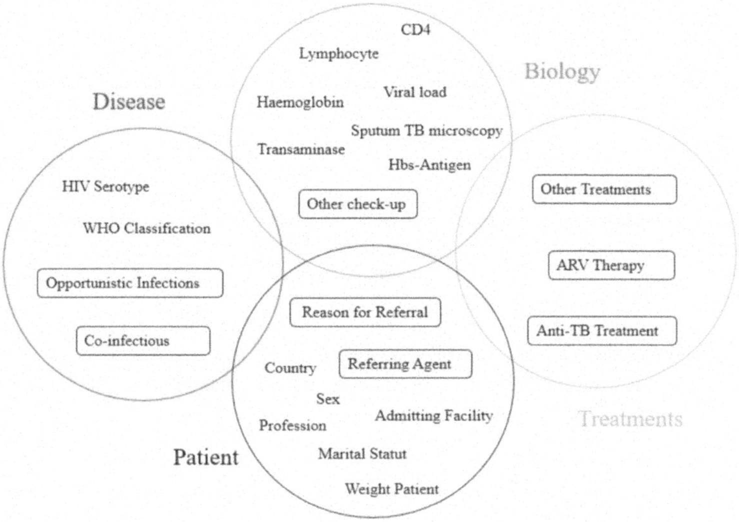

Fig. 1. The HIV/AIDS concepts and subdomains for referring and self-referring a patient.

- **Patient**, it's a physical person benefiting medical examinations and treatment provided by a doctor or health professional to deal with illness or injury [8]. In the file, we see a set of concepts and relationships that describe the reference reasons, the treating doctor, the patient's information (sex, weight, marital status, etc.), the referral site, etc.

The sub-domains presented in Fig. 1 cover the HIV/AIDS domain ontology and will be able to meet our need to link PLHIV monitoring data, but not sufficient on their own to satisfy the two last needs which are to obtain an ontology of the process for monitoring PLHIV and then to describe and share the resources for monitoring PLHIV.

To do this, not only will we have new concepts in the HIV/AIDS domain ontology, but in addition to the four sub-domains (Biology, Disease, Treatment, and Patient), we need a final sub-domain which is:

- **Anatomy**, it's the scientific study of the structure of organisms, including their systems, organs and tissues. It analyses the morphology and position of the different parts of the body, as well as their relations between them [9]. According to [10], from the clinical manifestations of AIDS we can cite the organs (lungs, central nervous system, digestive tract, skin, haematopoietic system) that may be affected. Whereas if we speak of organs, we are referring to anatomy.

The Fig. 2 gives not only an overall representation of the ontology of the monitoring process for PLHIV but it also allows to make the description and sharing the resources for monitoring PLHIV, because it is the person living with HIV who is the central element and around him we have the others sub-domains.

In this section, we have expressed the need to build an HIV/AIDS domain ontology for monitoring PLHIV and the contours of that ontology. A state of the art of existing

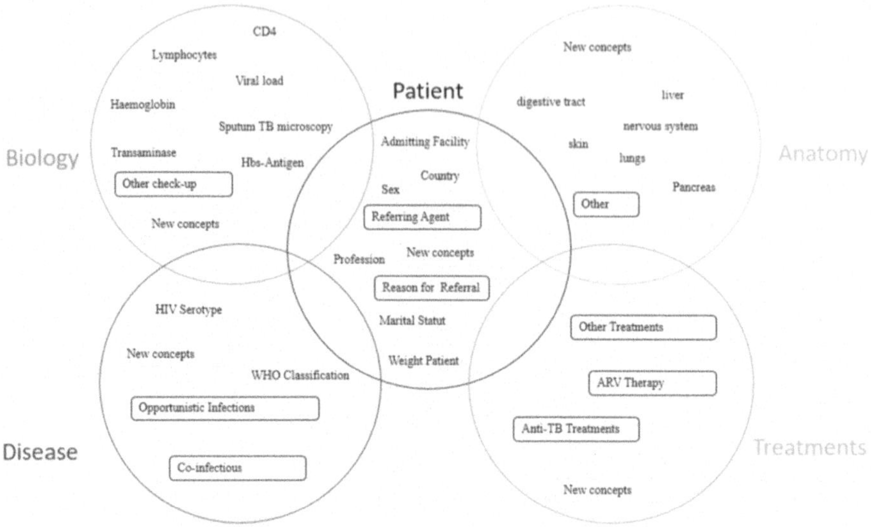

Fig. 2. HIV/AIDS sub-areas for the description of PLHIV follow-up processes.

terminology resources in the field of HIV/AIDS is required in order to choose and exploit these resources during construction.

These needs, presented, show an ontology that is not generalist but rather centred on the details of monitoring PLHIV for (1) link data from references and self-references of patient (2) describe the process of monitoring PLHIV in a process ontology and (3) describe and share follow-up resources when referencing.

3 State of the Art of HIV/AIDS Terminology Resources

In the remainder of this article, the term termino-ontological resources (TOR) will be used to represent those different types of resources that are terminological or ontological resources, which are built for a precise need. Our objective being to create an HIV/AIDS domain ontology for monitoring PLHIV, it is important to check the existing TOR that are related with the HIV/AIDS domain and that we will serve as a basis for the building of the ontology of this domain.

3.1 The Terminology Resources in Medicine

The medicine is one of the predominant fields for RTOs, and it's a fairly vast environment that contains a multitude of those resources, each of which responds to a different category of need. So, according to the cutting of our ontology into sub-domains, we will to present some biomedical ontologies that are related to those different sub-domains. Among those TOR, we can cite:

- Gene Ontology (GO) is the world's largest source of information on gene functions. It provides structured, controlled vocabularies and classifications of molecular and cellular biology [11]. This ontology contains almost all the concepts of biology, it takes in charge HIV/AIDS but with the exception of a few concepts for monitoring HIV/AIDS.

- Foundation Model of Anatomy (FMA) is a spatial-structural ontology of the entities and relations which together form the phenotypic structure of the human organism at all biologically salient levels of granularity [12]. It is an ontology that relates to the structural organisation of the human body, it holds several concepts for our anatomy sub-domain, but it also contains concepts that are not specific to the HIV/AIDS domain.
- Systematized Nomenclature of Medicine Clinical Terms (SNOMED CT) is a nomenclature with clinical terms, organized and structured, composed of concepts, descriptions, correlations, and categories. This is a multi-axial nomenclature which contains clinical data on health [13]. This general ontology covers all areas of healthcare, it holds many concepts for all our sub-domains, but is not specific to HIV/AIDS.
- Medical Subject Headings (MeSH) is a biomedical thesaurus that encompasses all the concepts appearing in medical literature, it's a tool powerful tool to carry out a targeted search [14]. It is used in particular for indexing and searching for medical information, but does not adequately represent the complex, cross-disciplinary nature of the disease at HIV/AIDS [16].

However, all of the cited above biomedical TOR are constructed to meet the diverse needs of scientific and medical organisations. That's why they contain the essential concepts to cover our different sub-domains, but most of the relationships between those concepts are not specific to the field of HIV/AIDS. So, in the following, we'll talk about terminological and ontological resources which are specific to HIV/AIDS.

3.2 The Terminology Resources on HIV/AIDS

The HIV/AIDS terminology resources contain the most commonly used terms that we encounter in the literature while speaking of HIV/AIDS. The terms are defined by a scientific community which activates in this field. Among these resources, we can cite:

- Encyclopaedic Dictionary of AIDS-Related Terminology a terminology that seeks to detail generally accepted meanings for the various words and phrases found in discussions of the interdisplinary aspects of the pandemic [15]. However, it should not be construed as a comprehensive listing of every term, but rather, it is a presentation of the more common verbiage included in discussions of HIV/AIDS.
- HIV/AIDS and HIV/AIDS-Related Terminology: A Means of Organizing the Body of Knowledge is presented as a tool to facilitate the organization and management of the body of knowledge concerning infection due to HIV and its end stage AIDS, and is intended to apply to works in all formats [16]. However, to cope with the daily evolution of the field of HIV/AIDS, the authors have reserved parts or persons who use this terminology to add terms in the appropriate place.
- UNAIDS Terminology Guidelines is implemented at United Nations Programme on HIV/AIDS (UNAIDS) to strengthen the global response to the AIDS epidemic. Those guidelines are freely available for use by staff members, colleagues in the Joint

Programme of the 11 Cosponsoring organizations and other partners working in the global response to HIV [17].
- Glossary of hiv/aids-related terms included in the Glossary represent the vocabulary of HIV used in a variety of federal resources, including approved HIV/AIDS medical practice guidelines and published results from HIV-related research studies [18].

The HIV terminology resources are flagship documents in the HIV field because they have been obtained on consensus of experts from the field. However, there resources focus on the study of terms of a language. So if we take a closer look at those resources, on one hand they have many more terms in our diseases and treatments subdomains, on the other hand, they have less terms for our other subdomains. Indeed, they are potential candidates for the construction of our HIV/AIDS domain ontology for monitoring PLHIV.

3.3 The Ontological Resources on the HIV/AIDS

These ontological resources were designed for a specific purpose in relation to the field of HIV/AIDS. So, as we share the same domain but do not have the same needs, it is important to carry out a literature review of these resources. During our research we were able to find these ontologies, which we describe below.

- HIV/AIDS ontology an ontology that enables a system to answer all the questions that a person can ask about HIV/AIDS from their standard mobile phone [19]. Although, this ontology was created using text extraction methods based on a single reference booklet (the MASA[5]) it hasn't been validated by experts in the field.
- Mining Ontology is an ontology that enables a system called Chronominer to discover temporal associations between genetic mutations in the viral genome, drug regimens through the administration of a specific drug and the occurrence of poor clinical outcomes with one or more drugs [20]. This ontology specifies the hierarchical "is-a" relationship between information on drugs, mutations and viral load. Thus, in terms of relationships, this ontology is poor to cover all the aspects of HIV/AIDS.
- Ontology Anti-AIDS Inhibitor is an ontology that enables a system to extracts information about the anti-AIDS inhibitor from the biological literature to provide a more powerful information access tool to help researchers overcome the challenge of information overload [21]. So, this ontology is a taxonomy that doesn't have many concepts or relations to cover our subdomain biology.
- HIV-associated Entities in Research Ontology (HERO) is an ontology that provides common data elements, relationships, correspondences and a general knowledge base for clinical and behavioural research associated with HIV [22]. However, the outline of this ontology is limited to clinical, behavioural and biomedical research on HIV. That is why it is difficult and even impossible to have concepts on biology, on the opportunistic and co-infectious diseases of HIV/AIDS.

BioPortal Ontology is a Web portal which allows access to a library of biomedical ontologies and terminologies via OCBN (National Center for Biomedical Ontology)

---

[5] **MASA**: is booklet Antiretroviral Therapy booklet which consists of 205 most frequently asked questions About HIV/AIDS and ARV Therapy. The booklet was prepared by the Ministry of Health in Botswana.

Web services [23]. This portal allows several medical ontologies to be put online, and that's what allowed us to be able to search for ontologies with the keyword "HIV", and we have as result the list of the following ontologies:

- HIV Ontology (HIV)[6] is an ontology of HIV that is supposed to encompass all knowledge on the domain of HIV, according to its description the BioPortal Ontology.
- HIVOntologymain.owl (HIVO004)[7] is an ontology which groups together several other ontologies in order to enable piloting new research on HIV/AIDS or other viral diseases to be conducted.

But if we take a closer look at those ontologies of BioPortal Ontology (HIV, HIVO004), it reveals the absence of many concepts such as opportunistic infections, co-infections, etc.

The HIV/AIDS ontology resources have been designed for well-defined goals purposes and their content varies according to the type of ontology. Thus, the resources defined above were found in the literature and the common factor that unites them is that they are not available, except those of BioPortal Ontology (HIV, HIVO004) which are available but have no scientific reference that speaks of them and no sources have been found on how they were built.

4 Synthesis and Analysis of the State of the Art of HIV/AIDS TOR for Monitoring PLHIV

We have reviewed the state of the art of existing RTOs in the field of HIV/AIDS. There is a wide variety of TOR containing general medical knowledge for needs different, varied and which are sometimes far removed from our ontology requirement described in the first part of this document, which delimits the outline of our future ontology in the sub-domains: biology, disease, treatment, patient and anatomy; covered by some of these resources and not for others. So, thus, during our research we had to make the following observations:

On the one hand, we noted that the GO and FMA resources are ontologies that contain the essentials concepts for our biology and anatomy sub-domains, thus we note some inconsistencies, redundancies and absence of relations in these TOR for example the GO [24, 25] and the FMA [26, 27]. In the same way, the nomenclature SNOMED CT and the MeSH confirm their position as biomedical ontologies by holding several concepts for each sub-domain, they too have imperfections due to the absence of certain hierarchical relationships and concepts [28–30]. However, the Encyclopedic Dictionary of AIDS-Related Terminology et HIV/AIDS, the HIV/AIDS-Related Terminology: A Means of Organizing the Body of Knowledge, the UNAIDS Terminology Guidelines and the Glossary of hiv/aids-related terms have many terms in the disease and treatments sub-domains, but few terms for the biology, the patient and the anatomy sub-domain. In fact, all these resources were created for a very specific need which is different from ours, that is why we have drawn up a table (see Table 1) summarising those different TOR in which we try to see what exists in relation to our sub-domains.

[6] **HIV Ontology (HIV):** https://bioportal.bioontology.org/ontologies/HIV.

[7] **HIVOntologymain.owl (HIVO004):** https://bioportal.bioontology.org/ontologies/HIVO004.

Table 1. Representations of the RTO according to our sub-domains

TOR	HIV/AIDS subdomains				
	Biology	Disease	Treatments	Patient	Anatomy
GO	+	o	o	o	+
FMA	+	o	o	o	+
SNOMED CT	+	+	+	+	+
MeSH	+	+	+	+	+
Encyclopedic Dictionary of AIDS-Related Terminology	–	+	+	–	–
HIV/AIDS and HIV/AIDS-Related Terminology: A Means of Organizing the Body of Knowledge	–	+	+	–	–
UNAIDS Terminology Guidelines	o	–	–	–	o
Glossary of hiv/aids-related terms	–	+	+	–	–

+: many concepts (or terms); – : few concepts (or terms); o: no concepts (or terms)

On the other hand, we were faced with the unavailability of certain resources such as HIV/AIDS ontology, Mining Ontology, Ontology Anti-AIDS Inhibitor and HIV-associated Entities in Research Ontology (HERO), although they have scientific references attesting to their existence and the information's on how they were constructed. Although, the ontologies found in the BioPortal Ontology (HIV, HIVO004) are available, they have no scientific references, nor any information on how they were constructed. The following table provides a summary of these different TOR (Table 2).

Table 2. TOR representations of the following RTO: Availability, Scientific reference and Construction information

TOR	Available	Scientific reference	Construction information
HIV/AIDS ontology	NO	YES	YES
Mining Ontology	NO	YES	YES
Ontology Anti-AIDS Inhibitor	NO	YES	YES
HIV-associated Entities in Research Ontology (HERO)	NO	YES	YES
HIV Ontology	YES	NO	NO
HIVOntologymain.owl	YES	NO	NO

An analysis of these various collections shows that there are no RTOs to satisfy our need for an HIV/AIDS domain ontology for monitoring PLHIV in the cross-border zone between Senegal, the Gambia and Guinea-Bissau, and less the follow-up of PLHIV in a specific way. And this confirms what was said in [16], existing systems, such as not limited to HIV/AIDS-fail to represent the intricacies of the malady; while those that have been produced in response to the illness and its impact on various communities do not include specifics concerning biomedical aspects of HIV and associated diseases and infections. Consequently, the interest of our work to propose an ontology for the field of HIV/AIDS for the monitoring of PLHIV and for the needs defined previously is no longer. However, the reuse of several RTOs is envisaged because some of these resources contain some of the concepts for our different sub-domains. For example, seen the decomposition of the sub-domains defined in the Fig. 2, it is necessary to reuse the TOR in medicine such as: SNOMED CT and MeSH. Just as the reuse of terminological resources of HIV/AIDS is also necessary.

To exploit well all the TOR identified in the construction of our HIV/AIDS domain ontology for monitoring PLHIV, it is primordial to choose a good methodological approach to ontology construction. It is this methodological choice that will take into account the reuse and/or re-engineering of TOR, which will be the focus of the next part of this works.

5 Conclusion

After recalling the context of the need to have a domain ontology on HIV/AIDS, we were able to identify the contours of this ontology, which are the sub-domains described in the first part of this article. However, the literature review enabled us to discover that there are many TOR in the medical domain but few are specific to HIV disease; although none of these TOR alone allow us to achieve our need, but we can reuse them during the construction of our ontology because they have concepts that are specific to the HIV/AIDS domain. To better operate all the TOR identified in the construction of this ontology, it is indispensable to start from a methodological approach to ontology construction. This methodological choice must take into account the possibility of reusing and/or reengineering existing TOR in the domain under study. As perspectives, in the next works we will propose a state of the art on ontology construction methodologies, this which will enable us to choose the methodology best apt to the construction of our ontology.

References

1. Lalou, R., Piché, V.: Et Centre français sur la population et le développement, Migration et SIDA en Afrique de l'Ouest, un état des connaissances. Centre français sur la population et le développement (CEPED), Paris (1994)
2. Dieng, Y., Diop, I., Faye, Y., Malack, C.A.: A sub-regional information system for monitoring and managing PLHIV in cross-border areas between Gambia, Senegal and Guinea Bissau. In: 9th International Conference, Cairo, France: ACM Press, pp. 1-7 (2019). https://doi.org/10.1145/3361570.3361576
3. Diop, I., Dieng, Y., Faye, Y., Malack, C.A., Cisse, O., Diouf, B.: Decrease the number of patients lost to follow-up in the monitoring of PLHIV in cross-border areas between The

Gambia, Senegal and Guinea Bissau. In: Proceedings of the 9th International Conference on Information Systems and Technologies, pp. 1–6 (2019). https://doi.org/10.1145/3361570.3361596
4. Studer, R., Benjamins, V.R., Fensel, D.: Knowledge engineering: principles and methods. Data Knowl. Eng. **25**(1–2), 161–197 (1998). https://doi.org/10.1016/S0169-023X(97)00056-6
5. Qu'est-ce que la biologie? Département de sciences biologiques - Université de Montréal. https://bio.umontreal.ca/departement/biologie/ (consulté le 6 juillet 2023)
6. Maladie: Géoconfluences, décembre (2022). http://geoconfluences.enslyon.fr/glossaire/maladie. consulté le 6 juillet 2023
7. Traitement (médecine): Wikipédia. 13 avril 2023. Consulté le: 6 juillet 2023. [En ligne]. Disponible sur: https://fr.wikipedia.org/w/index.php?title=Traitement_(m%C3%A9decine)&oldid=203284 171
8. Patient: Wikipédia. 28 avril 2023. Consulté le: 6 juillet 2023. [En ligne]. Disponible sur: https://fr.wikipedia.org/w/index.php?title=Patient&oldid=203758229
9. Anatomie: Wikipédia. 14 mai 2023. Consulté le: 6 juillet 2023. [En ligne]. Disponible sur: https://fr.wikipedia.org/w/index.php?title=Anatomie&oldid=204232279
10. Bennai, G., Belebad, H., Fakih, T.: Infections opportunistes sur VIH. Thesis, 2015. Consulté le: 5 août 2023. [En ligne]. Disponible sur: http://dspace.univtlemcen.dz//handle/112/8427
11. Gene Ontology Consortium: The Gene Ontology (GO) database and informatics resource. Nucleic Acids Res. **32**(suppl_1), D258-D261 (2004). https://doi.org/10.1093/nar/gkh036
12. Rosse, C., Mejino, J.L.V.: The foundational model of anatomy ontology. In: Burger, A., Davidson, D., Baldock, R. (eds.) Anatomy Ontologies for Bioinformatics. Computational Biology, vol. 6, pp. 59–117. Springer, London (2008). https://doi.org/10.1007/978-1-84628-885-2_4
13. Shahpori, R., Doig, C.: Systematized nomenclature of medicine–clinical terms direction and its implications on critical care. J. Crit. Care **25**(2), 364-e1 (2010). https://doi.org/10.1016/j.jcrc.2009.08.008
14. Chang, A.A., Heskett, K.M., Davidson, T.M.: Searching the literature using medical subject headings versus text word with PubMed. Laryngoscope **116**(2), 336–340 (2006). https://doi.org/10.1097/01.mlg.0000195371.72887.a2
15. Huber, J.T., Gillaspy, M.L.: Encyclopedic Dictionary of AIDS-Related Terminology. Routledge, New York (2000). https://doi.org/10.4324/9780203047804
16. HIV/AIDS and HIV/AIDS-Related Terminology : A Means of Organizing the Body of Knowledge. Routledge (2013). https://doi.org/10.4324/9781315863191. 16. « Guide de terminologie de l'ONUSIDA – 2015 » HIV Justice Toolkit ». https://toolkit.hivjusticeworldwide.org/fr/resource/unaids-terminology-guidelines-2015/ (consulté le 4 août 2023)
17. J. U. N. P. on Hiv/Aids: UNAIDS terminology guidelines (2011). Consulté le: 19 août 2023. [En ligne]. Disponible sur: https://digitallibrary.un.org/record/3952164
18. Clinical Info and HIVinfo are services of the U.S. Department of Health, and Human Services (HHS) and is maintained by the Office of AIDS, Research (OAR), which is part of the National Institutes of Health, et (NIH)., Glossary of HIV/AIDS-Related Terms
19. Moeng, B., Ayalew, Y., Mosweunyane, G.: Experimental Evaluation of HIV/AIDS Ontology Construction Tools. présenté à 760: 761: 762: 763: The IASTED 2012 African Conferences, ACTA Press (2012). https://doi.org/10.2316/P.2012.763-007
20. Raj, R., O'Connor, M.J., Das, A.K.: An ontology-driven method for hierarchical mining of temporal patterns: application to HIV drug resistance research. AMIA Annu. Symp. Proc. **2007**, 614–619 (2007)
21. Zhang, C., Du, J., Zhang, R., Fan, X., Yuan, Y., Ning, T.: Extracting ınformation of anti-AIDS inhibitor from the biological literature based on ontology. In: Preparata, F.P., Fang, Q. (eds.) Frontiers in Algorithmics (FAW 2007). LNCS, vol. 4613, pp. 74–83. Springer, Heidelberg (2007). https://doi.org/10.1007/978-3-540-73814-5_7

22. Brown III, W.: Ontology-based semantic harmonization of HIV-associated common data elements for integration of diverse HIV research datasets. Columbia Universit (2016). https://doi.org/10.7916/D8Z89CQV
23. Whetzel, P.L., et al.: BioPortal: enhanced functionality via new Web services from the National Center for Biomedical Ontology to access and use ontologies in software applications. Nucleic Acids Res. **39**(suppl_2), W541-W545 (2011). https://doi.org/10.1093/nar/gkr469
24. Yu, A.C.: Methods in biomedical ontology. J. Biomed. Inform. **39**(3), 252–266 (2006). https://doi.org/10.1016/j.jbi.2005.11.006
25. Mougin, F.: Identifying redundant and missing relations in the gene ontology. Stud. Health Technol. Inform. **210**, 195–199 (2015)
26. Gu, H.H., Wei, D., Mejino, J.L., Jr., Elhanan, G.: Relationship auditing of the FMA ontology. J. Biomed. Inform. **42**(3), 550–557 (2009). https://doi.org/10.1016/j.jbi.2009.01.001
27. Luo, L., Mejino, J.L., Jr., Zhang, G.Q.: An analysis of FMA using structural self-bisimilarity. J. Biomed. Inform. **46**(3), 497–505 (2013). https://doi.org/10.1016/j.jbi.2013.03.005
28. Mortensen, J.M., et al.: Using the wisdom of the crowds to find critical errors in biomedical ontologies: a study of SNOMED CT. J. Am. Med. Inform. Assoc. **22**(3), 640–648 (2015). https://doi.org/10.1136/amiajnl-2014-002901
29. Cui, L., Zhu, W., Tao, S., Case, J.T., Bodenreider, O., Zhang, G.Q.: Mining non-lattice subgraphs for detecting missing hierarchical relations and concepts in SNOMED CT. J. Am. Med. Inform. Assoc. **24**(4), 788–798 (2017). https://doi.org/10.1093/jamia/ocw175
30. Varghese, J., Dugas, M.: Frequency analysis of medical concepts in clinical trials and their coverage in MeSH and SNOMED-CT. Methods Inf. Med. **54**(01), 83–92 (2015). https://doi.org/10.3414/ME14-01-0046
31. Formulaire de reference: https://crossborder-platform.org/index.php/formulaire-de-reference/ (consulté le 27 août 2023)

Smart and Green Industry 4.0: A Review

Khaled Hamdan(✉), Okba Kazar, Nadeen Zaki, Aysha Al Ahbabi, Mariam Alkaabi, Sarah A. L. Haithami, and Radwan Hamdan

CIT, UAEU, Al Ain, UAE
{Khamdan,o.kazar,700037065,202100412,202003085,201950178, 201950066}@uaeu.ac.ae

Abstract. The Fourth Industrial Revolution (Industry 4.0) represents a fundamental change in the manufacturing and production systems, presenting both opportunities and challenges for businesses. As environmental sustainability gains increasing attention, smart and green technologies have emerged as vital components of Industry 4.0, allowing businesses to optimize their operations while minimizing their environmental impact. The study investigates the design and implementation of smart and green Industry 4.0, focusing on key technologies, best practices, and case studies. The developed research in this paper investigates the intelligent and environmentally friendly aspects of Industry 4.0, examining valuable case studies across various industries and disciplines. The research concludes by providing valuable insights into optimal approaches for designing and implementing and Sustainable Industry 4.0 Solutions, as well as discussing future trends and directions in the field. The study aims to explore the design and implementation of smart and green Industry 4.0, investigate future trends and directions in this area, discuss case studies of successful implementations in different sectors and regions, and identify best practices for developing and implementing Industry 4.0.

Keywords: Industry 4.0 · smart and green manufacturing · sustainability · environmental impact

1 Introduction

The breakthrough of manufacturing has experienced a revolutionary change with the rise of Industry 4.0, forced by the incorporation of state-of-the-art digital advancements such as the Internet of Things (IoT), advanced big data analysis, and artificial intelligence (AI). These technologies have empowered businesses to enhance production, reduce costs, and improve overall operations. However, as environmental responsibility gains prominence, the fusing of smart and green technologies into Industry 4.0 becomes increasingly crucial. By incorporating green and intelligent technologies, businesses can effectively minimize carbon emissions, enhance energy conservation, and advocate responsible manufacturing methods. This raises the question of how the principles of Smart and Green Industry 4.0 can be effectively applied in the design and implementation of industrial processes, and what benefits can be achieved as a result. The specification of the components of the

M. R. Laouar et al. (Eds.): ICISAT 2023, LNNS 981, pp. 88–99, 2024.
https://doi.org/10.1007/978-3-031-60591-8_8

smart green industry is a key aspect to address. Traditional industrial processes have often been associated with negative environmental and social impacts, including pollution, waste, and unsafe working conditions. The emergence of the fourth industrial revolution, presents an opportunity to integrate advanced technologies and sustainability concepts into industrial processes, creating smart and green factories that are both efficient and environmentally responsible. However, designing and implementing such systems is a complex task that requires expertise in engineering and sustainability.

Therefore, the primary problem to be addressed is how to effectively design and implement Smart and Green Industry 4.0 systems to achieve sustainable, safe, and efficient industrial processes while ensuring economic profitability. The main objective of the research paper (Hassoun et al., 2022) is to investigate the potential effects of utilizing environmentally friendly approaches and integrating Industry 4.0 technologies in order to realize the sustainable development objectives within the realm of the food industry. The work described in the article may involve further research and development of these technologies, particularly in the context of the food industry, as well as a review of potential implementation challenges and corresponding solutions.

The research conducted by Umar et al. (2022) is to analyze the connection between Industry 4.0 and the adoption of green supply chain management (GSCM) practices. Additionally, the study aims to assess how these factors contribute to the financial and environmental outcomes. The report suggests leveraging digital technology to promote sustainable development and enhance supply chain management, procurement, and manufacturing. It also emphasizes the incorporation of green concepts into business strategies and the importance of regulatory agencies upholding clear environmental standards to achieve sustainable performance. Manufacturing firms are advised to fully leverage the capabilities offered by Industry 4.0 technologies as they contribute to overall sustainability performance.

The study by Nyangchak (2022) offers a comprehensive driving factors that drive the green sector to facilitate the transition to a net-zero emissions economy, thereby contributing to climate stabilization efforts. It is suggested a detailed framework encompassing different components to assess plans for achieving green industrial growth within a net-zero economy.

2 The Literature Review

The research carried out by (Tripathi et al., 2023) introduces an innovative agile model known as the intelligent lean-green approach, which aims to enhance sustainability in Industry 4.0. This approach combines agile methodology, lean manufacturing, and environmentally friendly practices. The authors stress the importance of addressing sustainability challenges within the context of Industry 4.0, which involves the integration of digital technologies and automation in industrial processes [1]. The intended model aims to optimize operational performance while enhancing factors such as the work environment, worker efficiency, environmental impact, logistics management, and resource utilization. It consists of five steps: Gemba walk, planning of the production management system, determination of standards, implementation of improvements, and evaluation of production enhancement. The model incorporates principles from lean and green manufacturing, focusing on the dedication of the management team, strategic preparation, and holistic production data to attain a sustainable operational performance.

The study explores prior work related to process enhancement, lean production, intelligent manufacturing, and environmentally conscious manufacturing, and the integration of lean, smart, and green approaches. It highlights the advantages of integrating these approaches, such as improved operational performance, employee morale, and production systems within the context of Industry 4.0. The study also provides a comparison of key points discussed in the literature review, including the importance of waste elimination and the positive impact of integrating smart lean-green practices on operational performance. The proposed model is validated through two case studies conducted in different industry settings, demonstrating improvements in both operational and environmental performance. The study emphasizes the significance of the smart lean-green approach in optimizing production processes, enhancing sustainability, and achieving the goals of Industry 4.0. It suggests that future research can further explore the implementation of lean, green, and smart manufacturing models to promote long-term sustainability within the context of Industry 4.0 [1]. It suggests that future research can further explore the implementation of lean, green, and smart manufacturing models for long-term sustainability in Industry 4.0.

The research conducted by (Javaid, M., et al., 2022) explores various applications of Sustainability 4.0 in the manufacturing industry. These applications include smart manufacturing, integration of renewable energy, circular economy practices, optimization of the supply chain, energy and resource management, and life cycle assessment. The research highlights how these applications contribute to sustainable practices, improved efficiency, and cost reduction in manufacturing [2].

Moreover, the article emphasizes the significance of Sustainability 4.0 in enhancing sustainable manufacturing, addressing climate change, transforming entire systems and networks, promoting digitization, reducing waste, and improving employee safety. It emphasizes the potential of Sustainability 4.0 in reshaping both the economy and society, fostering inclusiveness and environmental sustainability.

Sustainability 4.0 possesses the transformative potential to combine digital technologies with sustainable practices to create manufacturing processes that are more efficient and environmentally friendly. The article emphasizes the importance of Sustainability 4.0 in achieving long-term operational sustainability by ensuring resources are not depleted faster than they can be replenished. The integration of sustainability and digitization holds promise in tackling environmental, social, and economic challenges within the manufacturing sector [2].

The research work (Nagasawa, J.T., et al., 2017) emphasizes the potential of Industry 4.0 in promoting clean energy technology and advancing sustainable development goals. It explores how concepts such as the IoT, artificial intelligence, robotics, and Big Data can expedite the manufacturing, distribution, and consumption of renewable energy [3].

The integration of Industry 4.0 technologies in the manufacturing sector can result in enhanced production efficiency, cost reduction, and higher-quality renewable energy products. Additionally, the use of digital technologies enables better energy management, demand response, and optimization of energy distribution through smart grids. Predictive maintenance and condition monitoring of renewable energy assets can increase their reliability and lifespan.

The crucial role of the United Nations Industrial Development Organization (UNIDO) lies in its promotion of inclusive and sustainable industrial development, closely aligned with the aspirations of Sustainable Development Goal 9 (SDG 9). SDG 9 aims to establish resilient infrastructure, foster innovation, and promote inclusive and sustainable industrialization. There are challenges to overcome in fully harnessing the potential of Industry 4.0 for clean energy. These challenges include upfront costs, technological integration, workforce adaptation, data security, regulatory frameworks, digital divide, and sustainability considerations [3].

The article "Sustainability Green Industry 4.0" introduces the SGI 4.0 framework, which focuses on the connection between sustainability, Industry 4.0, and environmentally friendly practices. The framework emphasizes the role of green processes in achieving sustainability goals through the integration of Industry 4.0 technologies. It provides comprehensive guidelines for assessing sustainability in this context.

While existing literature often associates Industry 4.0 with sustainability, the article highlights that the significance of green processes is frequently overlooked. The SGI 4.0 framework provides a comprehensive explanation of how green processes and Industry 4.0 collaborate to achieve sustainability goals. The authors developed the SGI 4.0 conceptual framework, which covers various processes enabled by Industry 4.0 and green technology to achieve sustainable results. The article also delves into the historical background of green procedures and technology, their significance for emerging sectors, and their positive effects on budgetary management.

In the context of Industry 4.0, the SGI 4.0 framework strives aims to offer tangible evidence of the sustainability results yielded of environmentally- green processes. It offers guidelines for assessing and evaluating sustainability in this setting, emphasizing the need for concrete proof of sustainability consequences.

In (Bednar, J., et al., 2020) "Industry 4.0: Current Status and Future Trends" the authors discuss a new industrial model characterized by intelligent, virtual, and digital performance. Intelligent manufacturing, also known as smart manufacturing, is explored as a model that optimizes production and processes through data utilization. The study emphasizes the necessity of the Industrial Internet of Things (IIoT) architecture platform in facilitating secure data transfer between machines and factories. It specifically focuses on enabling brownfield equipment with IIoT capabilities and performing data analytics at the edge and in the cloud [4].

The article concludes by mentioning that the book is organized into two sections: reviews and case studies. It serves as a valuable reference on Industry 4.0, including case studies covering a range of topics. Written by expert researchers, the book explores the concept of Industry 4.0 and its potential impact on future societies, including the transition towards Society 5.0.

Industry 4.0 is highly relevant and important in the real world, representing the digital transformation of industries with advanced technologies such as AI, robotics and Internet of Thing (IoT). Implementing Industry 4.0 can improve productivity, efficiency, and competitiveness while enabling customization and sustainable practices. Its impact extends to various sectors, and governments and organizations are investing in its development.

Industry 4.0 is shaping the future of industries, driving technological advancements, economic growth, and sustainability. The adoption of Industry 4.0 technologies can lead to improved productivity, efficiency, and competitiveness for businesses by enabling the real-time collection and analysis of data, as well as informed decision-making, this functionality serves to optimize processes, decrease expenses, and improve the quality of products. Industry 4.0 also offers opportunities for customization and personalized production, allowing businesses to meet changing customer demands and preferences [5].

The research (Bradu, P., et al.,2022) highlights the importance of Green Technology and Industry 4.0 in addressing global environmental challenges and achieving sustainable development. The paper emphasizes the need for a transition towards green and sustainable practices confronting the impacts of both climate change and the enduring global health crisis and to develop eco-friendly products, monitoring environmental conditions, and revolutionizing manufacturing processes using renewable resources and recyclable materials. The review also explores the role of nanotechnology, waste management, renewable energy, and bio-based materials in sustainable development. It emphasizes the importance of green economies, the post-COVID-19 transition towards sustainability, and the role of individuals, communities, industries, and governments in adopting green practices.

The paper provides valuable insights into the integration these technologies for a greener and more sustainable future, aligned with the United Nations' Sustainable Development Goals [6]. The study provides a concise overview full swing of green technology and Industrial Revolution 4.0 and their importance for a sustainable future. Prior to conducting their study, the authors employed a literature review strategy to collect insights about a range of green innovative technologies and IoT advancements. These technologies have the potential to be utilized in the creation of environmentally-friendly, robust, green, and eco-conscious products. It explores various green technologies and their potential in addressing global warming, promoting environmental biotechnology and green chemical engineering, and developing advanced biodegradable materials. It also provides a comprehensive overview of the relationship between green technology, Industry 4.0, and sustainability, covering aspects such as environmental biotechnology, green chemical engineering, product engineering, and environmental economies.

The project integrated diverse topics such as global warming, environmental biotechnology, circular economy, green design and manufacturing, and the Green New Deal, resulting in a unified and coherent presentation. It emphasizes the role of legislation, education, and motivation in promoting the adoption of green technologies and achieving sustainable transitions. The practical implications of green technologies and clean technologies are also discussed, highlighting their benefits for businesses and consumers. While the paper does not explicitly mention any limitations, it acknowledges the challenges and obstacles involved in implementing sustainability measures. Additionally, it recognizes the complexity of environmental biotechnology and the interdisciplinary nature of research in sustainable chemical engineering and advanced materials. In conclusion, the paper underscores the importance of creating a sustainable future by addressing climate change, embracing green technologies, and adopting eco-friendly practices at various levels, including individual, community, industry, and global [6]. The research

(Olah, J., et al., 2020), highlights the importance of industrialization in production and the adoption of Industry 4.0 by various industries, the remarkable level of product customization achieved stands out. Nevertheless, the introduction of Industry 4.0 has resulted in significant utilization of resources, raw materials, information, and energy, raising concerns about its environmental sustainability [7].

The study examining the impacts of Industry 4.0 on environmental sustainability and offering solutions for current environmental issues through the utilization of novel technologies is essential. Nonetheless, the integration of Industry 4.0 with the targets of sustainable development has the potential to significantly amplify the promotion of environmental sustainability. The paper emphasizes the need for policies to address environmental challenges and promote the adoption of new technologies. It suggests that Industry 4.0, when combined with sustainable practices, can create a sustainable framework that balances economic, environmental, and social aspects. The deployment scenario focuses on automation, which can lead to improved working conditions, reduced pollution, and lower production costs. The operation scenario emphasizes quality improvement, minimizing errors, and optimizing resource use. The goal of linking Industry 4.0 with sustainable development is to match it with objectives like clean energy, responsible consumption, and addressing climate change. The long-run scenario envisions a sustainable future through the adoption of well-designed policies.

However, the study acknowledges limitations and calls for further research into the social and economic sustainability aspects of Industry 4.0. It emphasizes the need for policymakers to stay updated on the evolving challenges and impacts of the changes that Industry 4.0 brings to environmental sustainability. The work highlights both the negative and positive impacts of Industry 4.0 on environmental sustainability. It emphasizes the importance of adopting sustainable practices, integrating with sustainable development goals, and implementing effective policies to ensure a sustainable future [7]. Limitations of this paper include the need for further research on the complex and important topic, considering environmental, social, and economic sustainability, as well as potential missed factors that could impact decision-making. The study highlights the significance of examining the structural, functional, and qualitative aspects, emphasizing the value of scenario-based research to anticipate impacts caused by Industry 4.0 and the importance of frequent studies to mitigate negative effects on environmental sustainability.

The research by (Zoubek et al., 2020) explores the idea of the Fourth Industrial Revolution, where important technologies merge to enable communication between machines and products. This era is defined by information technologies and intelligent systems that automate production, replacing human labor and requiring modern workplaces. The study focuses on environmentally mindful manufacturing during this revolution, presenting a model to evaluate environmental aspects of manufacturing within an industry [8].

The study concludes that most existing Industry 4.0 models overlook the environmental facets of manufacturing and their integration with the carbon footprint, a key measure of environmentally responsible production. Consequently, the authors propose an environmentally oriented maturity model that aligns with Industry 4.0 technologies employed by companies. The most noteworthy outcome of their study is the incorporation of an "environmental aspect" into the maturity model, tailored to suit Industry

4.0 technologies. Ultimately, the paper offers a structured framework for evaluating the environmental attributes of manufacturing processes within companies, highlighting the crucial role of ecologically responsible production in the era of the Fourth Industrial Revolution.

The article gives a thorough Industry 4.0 maturity model created specially to evaluate the environmental characteristics of manufacturing firms. This model intends to assess the level of environmental performance attained by businesses as they embrace and execute Industry 4.0 technologies and practices considering the growing emphasis to ecological sustainability. The significance of environmental sustainability in the industrial sector and the potential of Industry 4.0 to support sustainable practices are covered by the writers in the opening paragraphs. In the context of Industry 4.0, they emphasize the necessity for a standardized evaluation methodology that can efficiently monitor and benchmark the environmental features of industrial enterprises. Each level or stage in the suggested maturity model corresponds to a distinct level of environmental sustainability. Specific criteria, indicators, and performance metrics pertaining to the reduction of environmental impact, resource efficiency, waste management, energy consumption, and other pertinent environmental factors characterize each level. The article lists the primary factors and standards considered by the maturity model.

Energy management, carbon footprint, water use, waste management, recycling, eco-design, adoption of green technologies, and environmental certifications are a few examples of these. The model also considers how data-driven strategies and the integration of digital technology might improve environmental performance. Some of the limitations might include that the model may be difficult to implement, subjectivity in evaluation could lead to discrepancies, and the model's use might be restricted to the manufacturing industry without empirical support. The authors suggest a structured evaluation procedure to determine a manufacturing company's maturity level. This involves gathering relevant data on environmental characteristics, doing self-evaluation based on predetermined standards for each level, and comparing the outcomes to ascertain the current level of environmental sustainability. This enables businesses to pinpoint their areas for development and create goals to improve their environmental performance [8]. The paper (Li, X., et al.,2021) addresses the concept of greentelligent manufacturing (GIM) as an integrated paradigm for achieving sustainability in manufacturing. The study emphasizes the need to oppose the issue of global climate change and decrease greenhouse gas emissions in the manufacturing sector. It suggests combining smart technologies and eco-friendly green approaches to attain enhanced management, efficiency, and ecological responsibility within manufacturing systems. The study outlines the evolution and framework of GIM, highlighting enabling mechanisms such as changing manufacturing processes, reducing production waste, improving energy efficiency, and making manufacturing sustainable. It also discusses supporting tools and methods, including life cycle assessment, carbon pricing strategies, and carbon capture and storage. The study concludes that greentelligent manufacturing can play a crucial role in transitioning to a greener future while maintaining competitiveness in the industry [9]. The paper mainly focuses on the technical aspects of achieving green objectives in manufacturing and does not discuss the economic or social implications of implementing such technologies.

Combating global warming has become one of their most significant challenges as a result of global industrialization and economic growth. In order to increase supervision, efficiency, and sustainability in manufacturing processes, the article offers the idea of "Greentelligent Manufacturing" (GIM), which combines intelligent technology and environmentally friendly practices. It investigates the resources and processes that encourage and promote greener manufacturing and provides a framework for GIM. The three stages of emerging, transitional, and synergetic are used to explain the development of greentelligent manufacturing. It highlights many approaches and tools that may be used to improve the ecological the continuing viability of manufacturing procedures, like altering procedures, applying cutting-edge manufacturing techniques, utilizing eco-friendly materials, and employing creative product design. Additionally, it emphasizes the significance of waste elimination via additive manufacturing, selective automation, and lean manufacturing. Measures including machinery and industrial design, Internet of Things (IoT) technology, intelligent scheduling, and artificial intelligence are addressed as methods to boost energy consumption. It also discusses reconfigurable manufacturing, service-oriented manufacturing (SOM), and remanufacturing as sustainable manufacturing techniques. As essential elements of greentelligent manufacturing, supporting tools and techniques including life cycle assessment (LCA), carbon pricing mechanisms, and digital supply chain management are also included. To prioritize green technological developments and to facilitate a more seamless transition to a greener future, the paper proposes greentelligent manufacturing as a revolutionary interconnected manufacturing paradigm for Industry 4.0 and Industry 5.0. It offers several recommendations to lessen the negative effects of production on the environment, including the use of renewable energy sources, recycling and the circular economy, stricter environmental regulations, and promoting consumer awareness. The industrial sector can drastically minimize its environmental impact and contribute to development a sustainable future through proactively putting these steps into practice and constantly pursuing innovation [9].

The study conducted by (Chen & Huo, 2021) exposes that the green innovation efficiency within the Greater Bay Area (GBA) is increasing. It also identifies a specific spatial correlation among carbon emissions, albeit with a generally low overall correlation. The relationship between effective green innovation and carbon emissions demonstrates a pattern resembling an 'inverted U,' signifying an ideal point of green innovation efficiency that impacts emissions. The research recommends implementing measures to reduce carbon emissions within a practical scope, while also pinpointing avenues for future investigation and enhancement.

This study enhances our comprehension of how green innovation efficiency links to decreasing carbon emissions in urban clusters. Moreover, it offers valuable insights for shaping policy recommendations [10]. This study aims to enhance the effectiveness of green innovation and the reduction of carbon emissions within urban clusters, with a specific emphasis on the Guangdong-Hong Kong-Macao, GBA. The results indicate a nonlinear connection between green innovation efficiency and the intensity of carbon emissions, revealing a distinct threshold. The article provides policy recommendations, such as strengthening cooperation between cities, formulating tailored policies based on local conditions, improving the innovation transformation platform, and enhancing the extent of eco-friendly green innovation in urban clusters. However, the research acknowledges limitations in sample size, data availability, and the need for further research on influencing factors and spatial correlations. Future studies should aim to overcome these limitations and expand the research to include a broader range of regions and longer time periods to enhance the reliability and generalizability of the findings [10].

Environmental degradation and climate change are among the most pressing challenges globally in the 21 century. The present climate change forecasting still contains many unknowns. The Studies indicate that human actions elevated atmospheric carbon dioxide (CO2) levels from pre-industrial revolution levels to a point in the early 1990s. Subsequently, the average annual temperature of the Earth's surface has risen by 0.6 °C over the last century. Parties held climate discussions in Bonn, Germany, in 2018, obtaining the agreement's specifics inside the UNFCCC framework. Limiting carbon emissions is aided by environmental regulations, green innovation, risk indices, and renewable energy studies. Because carbon emissions have an externality, governments have passed rules and regulations to encourage emission reduction. As nations desire fewer limitations, more substantial rewards, and accountability, achieving climate progress poses difficulties. In 2020, China surpassed global commitments and curbed rapid growth, emitting 9.894 billion tons of CO2. This marked a reduction in carbon intensity by 18.8% since 2015. The GBA in Southern China, spanning Guangdong Province along with Hong Kong and Macao, is an economically dynamic region comprising nine Guangdong cities alongside the two special administrative regions. The GBA has acknowledged the significance of green innovation efficiency in decreasing carbon emissions and supporting sustainable development as concerns about climate change and environmental sustainability grow. The GBA has actively promoted green innovation efficiency, driven by several factors [10]:

1. Strong Government Support: The region benefits from robust government support, which creates a conductive environment for green innovation initiatives.
2. Technological Advancements: Technological progress has played a significant role in enabling and accelerating green innovation efforts in the GBA.
3. Collaboration and Knowledge Sharing: The GBA emphasizes collaboration and knowledge sharing among stakeholders to foster innovation and address climate change and environmental sustainability concerns.

3 Summary of the Previous Presented Papers

In this section we will summarize the most important characteristics of the previous studied works on Smart and Green Industry 4.0. The table below (Table 1) contains the comparative study.

Table 1. Comparative study

Ref.	Approach	Strength	Limitations
Tripathi et al. [1]	Smart Lean-Green Approach integrating Agile, Lean, and green principles to improve sustainability in Industry 4.0	Novel model integrating agile, lean, and green. Comprehensive model covering operational and environmental aspects	No specific implementation challenges discussed
Javaid, M. et al. [2]	Investigation of viability applications in manufacturing are 4.0. Promoting circular economy, sustainable energy, and other concepts	Diverse applications and practical focus and improvements	Limited discussion on drawbacks or limitations of Sustainability 4.0
UNIDO [3]	UNIDO's role in sustainable industrial development (SDG 9)	Clear articulation of UNIDO's role. In sustainable industrialization. Identification of challenges in adopt Industry 4.0	Focus on UNIDO's role, lacks specific approaches
Pech, M., et al. [4]	Introduction of SGI 4.0 framework for sustainability assessment	Comprehensive framework for Industry 4.0	Lacks real-world case studies for validation
Ortiz, J. H., et al. [5]	Discussion of intelligent manufacturing and IIoT in Industry 4.0	Comprehensive overview and case studies and its potential impact	Limited discussion on challenges of Industry 4.0
Bradu, P., et al. [6]	Integration of green technology and IR 4.0 for sustainable development	Thorough exploration of green technology. Addresses aspects of sustainable development	No explicit limitations mentioned. Lack of specific implementation strategies or case studies

(*continued*)

Table 1. (*continued*)

Ref.	Approach	Strength	Limitations
Oláh, J., et al. [7]	Relationship between sustainability, Industry 4.0, and green technology and the pursuit of sustainability are all intertwined	In-depth discussion of relationship between green technology and IR 4.0	No explicit limitations but calls for research
Zoubek, et al. [8]	The Fourth Industrial Revolution affects Sustainable Manufacturing Practices	Development of an Environmental Attributes Maturity Model	Focus on assessing attributes, limited impact
Li, X., et al. [9]	Introduction of intelligence manufacturing (GIM) for sustainability	In-depth exploration of GIM concept. Highlights supporting tools and mechanisms	Limited discussion on economic and social implications of GIM
Chen, L., & Huo, C. [10]	Examination of the Efficiency of Green Innovation in Reducing Carbon Emissions	Quantitative Exploration of the Correlation Between Green Innovation and Carbon Emissions	Restricted Exploration of Broader Global Implications

4 Conclusion

In this article we have established a bibliographical synthesis concerning the works relating to the smart and green industry. Smart and green technologies have emphasized the significance of smart and green technologies in Industry 4.0, enabling businesses to optimize their operations while minimizing their environmental footprint. A noteworthy outcome of this work is the integration of the "environmental aspect" integration into the developmental stages of Industry 4.0 technologies. This draws attention to the environmentally friendly aspects of manufacturing processes and emphasizes the value of eco-conscious production in the context of the Fourth Industrial Revolution. As a perspective to this research work, we will develop smart and green industry architecture using distributed artificial intelligence model based on multiagent system and IoT.

Acknowledgement. This work was supported by fund from the SURE plus research program from UAEU, under Grant Code: 2842.

References

1. Tripathi, V., et al.: An innovative agile model of smart lean-green approach for sustainability enhancement in industry 4.0. Sustainability **15**(6), 3456–3478 (2023)

2. Javaid, M., Haleem, A., Singh, R.P., Khan, S., Suman, R.: Sustainability 4.0 and its applications in the field of manufacturing. Sustainability **13**(12), 6752 (2022)
3. Nagasawa, T., et al.: Accelerating clean energy through Industry 4.0: manufacturing the next revolution. UNIDO, Vienna (2017)
4. Pech, M., Vrchota, J., Rolnek, L., Bednar, J.: Sustainability green ındustry 4.0. In Encyclopedia (2020)
5. Ortiz, J.H., Marroquin, W., Cifuentes, G.: Industry 4.0: current status and future trends (2020)
6. Bradu, P., Biswas, A., Nair, C., Sreevalsakumar, S., Patil, M., Kannampuzha, S., Gopalakrishnan, A.V.: Recent advances in green technology and industrial revolution 4.0 for a sustainable future. Environ. Sci. Pollut. Res. **30**, 1–32 (2022)
7. Oláh, J., Aburumman, N., Popp, J., Khan, M.A., Haddad, H., Kitukutha, N.: Impact of industry 4.0 on environmental sustainability. Sustainability **12**(11), 4674 (2020)
8. Zoubek, M., Poor, P., Broum, T., Basl, J., Simon, M.: Industry 4.0 maturity model assessing environmental attributes of manufacturing company. Appl. Sci. **11**(11), 5151 (2021)
9. Li, X., Wang, B., Peng, T., et al.: Greentelligence: smart manufacturing for a greener future. Chin. J. Mech. Eng. **34**, 116 (2021)
10. Chen, L., Huo, C.: Impact of green innovation efficiency on carbon emission reduction in the Guangdong-Hong Kong-Macao GBA. Sustainability. **13**(23), 13450 (2021)
11. Hassoun, A., Prieto, M.A., Carpena, M., Bouzembrak, Y., Marvin, H.J., Pallares, N., Bono, G.: Exploring the role of green and Industry 4.0 technologies in achieving sustainable development goals in food sectors. Food Res. Int. **162**, 112068 (2022)
12. Umar, M., Khan, R., Yusliza, M., Ali, S., Yu, Z.: Industry 4.0 and green supply chain practices: an empirical study. Int. J. Prod. Perform. Manage. **71**(3), 814–832 (2022)
13. Nyangchak, N.: Emerging green industry toward net-zero economy: a systematic review. J. Clean. Prod. **378**, 134622 (2022)

Machine Learning Algorithms Based Models for Forest Fires Prediction: An Analytic and Comparison Study

Faroudja Abid(✉)

Microelectronic and Nanotechnology Division, Centre de Développement des Technologies Avancées, Algiers, Algeria
fabid@cdta.dz

Abstract. This paper describes the outcomes of an investigation of several machine learning algorithms as models for forest fires prediction. The conducted comparison and analytic study is based on the performance evaluation of the employed algorithms in terms of accuracy, recall and precision. The aim of this study being the identification of the most efficient ML algorithm based model for forest fires prediction. We have considered in this study four datasets. The prediction is based on the meteorological data corresponding to the critical weather elements that influence the forest fires occurrence (namely temperature, relative humidity, wind speed and rain) and the fire weather index system components. This study revealed that the naive Bayesian and the decision tree algorithms perform well for all datasets, thus they are suitable for our purpose since they give tradeoff between performance (in terms of accuracy, recall and precision) and the model build time. This research provides an orientation for the selection of the most suitable ML algorithm as model for forest fires prediction in Algeria.

Keywords: Algerian forest fires dataset · Fire prediction · Machine learning.

1 Introduction

Forest fires cause negative consequences for the environment, economy and human lives. The fires prediction basically, deals with forecasting the forest fires outbreak probability before its initial ignition. The prediction of forest fires is important for preventing their occurrence. In recent years, machine learning (ML) algorithms, have been widely investigated for forest fires detection and prediction. In fact, the ML algorithms are applied in almost all fields for the automation purpose in several applications including the forest fires prediction and detection. Almost all types of machine learning algorithms are exploited for this issue [1].

A lot of researches have been carried out for fire occurrence modeling using the ML algorithms based models [2–9]. Several ML algorithms based fire detection models applied to wireless sensor network were compared in [2]. The study exposed in [3] investigated the use of ML algorithms for forest fires occurrence prediction in China.

M. R. Laouar et al. (Eds.): ICISAT 2023, LNNS 981, pp. 100–110, 2024.
https://doi.org/10.1007/978-3-031-60591-8_9

Four ML algorithms have been considered in this study, namely the Artificial Neural Network (ANN), the Radial Basis Function network (RBF), the Support-Vector Machine (SVM), and the Random Forest (RF). The used data includes fire hotspots, meteorological conditions, terrain, vegetation, and socioeconomic. Yongqi et al. reported results in terms of accuracies that range from 75.8% to 89.2% for the four forest fires prediction models. According to this study, the highest accuracy was obtained for the random forest algorithm. They also identified the meteorological, topographical, human and vegetation factors as the main driving factors of forest fires occurrence in China. In [4] the authors adopted a Naive Bayes classifier as a model for forest fires detection, the reported precision was about 94%. A forest fires detection system based on the Fire Weather Index (FWI) system components (the Fine Fuel Moisture Code (FFMC) and the FWI), was exposed in [5].

The logistic regression and the decision tree (DT) and its ensembles were investigated in [7] for the evaluation of the forest fires occurrence in Slovenia. This study revealed that the bagging DT performs better. The reported accuracies were about 81%, 84.9% and 84.4% for the simple DT, the bagging DT and the boosting DT, respectively. The study reported in [8] revealed that the SVM classifier based solution for burned area prediction is suitable for small fires detection rather than large fires.

In our previous work [9], we have presented a case study of the decision tree algorithm for predicting forest fires in Algeria using a dataset that includes meteorological conditions (the first dataset). In the present work we have extended the study for several ML algorithms as models for forest fires prediction, and the evaluation is done using four datasets. We discuss in this work the application of several ML algorithms based models for forest fires prediction in Algeria, which is one of the most affected countries by this disaster [10]. We conduct an evaluation and comparison of several ML algorithms based models using three performance metrics, namely, accuracy, recall and precision. The aim of this study being the identification of the most efficient ML algorithm for forest fires prediction.

Our previous review paper [1] revealed that the mostly applied ML algorithms for forest fires modeling are the ANN [11], the logistic regression [12], and the decision tree [13] and its ensembles, mainly the Random Forest (RF) [14]. We investigate in the present study other ML algorithms such as the Bayesian (the BayeNet, and the Naive Bayesian (NB)), the SVM [15] and the K-Nearest Neighbors (KNN). The used dataset is the publicly available Algerian forest fire dataset [16]. This dataset includes the critical weather elements influencing the forest fires occurrence, namely temperature, relative humidity, wind speed and rain. The dataset also includes the FWI system components (related to the fuel content and moisture content). The forest fires dataset instances are classified as "fire" and "not fire" classes using the Canadian FWI system for fire danger rating. More details about this dataset will be given in Sect. 2.2.

This paper is organized as follows: In Sect. 2, we present the adopted method for this investigation, including the algorithms selection and the employed datasets. Section 3 deals with the results analysis and discussion, we mainly, present the applied ML algorithms performance evaluation for the identification of the suitable ML algorithm for forest fires prediction modeling. Finally, Sect. 4 concludes this paper.

2 Method

In this section, we expose the ML algorithms selection and the description of the constructed dataset, which is used for the ML based forest fires prediction models performance evaluation. We also present the performance metrics considered for the applied ML algorithms performance evaluation.

2.1 Machine Learning Algorithms Selection

The applied ML algorithms selection is based on our previous work [1]. According to this survey, the most employed algorithms were the ANN (mainly the Multi-Layer Perceptron (MLP)), the logistic regression and the DT and its ensembles. Other types of ML algorithms have been investigated, such as the SVM, the Bayesian and the fuzzy logic. However, these models are less employed in comparison to the ANN and logistic regression models. It is noteworthy to precise that the deep ANN (DNN) (mainly the convolutional ANN (CNN)) gives very acceptable accuracies when explored for fire modeling. In fact, we did not investigate the deep learning algorithms (such as the CNN) in this study, since the DNNs are more suitable for large datasets. In addition to these ML algorithms, we have also investigated in this comparison study, the KNN, the Meta algorithms that belong to the field of ensembles learning such as the Bagging, the Boosting and the random forest algorithms.

2.2 Software Tool/Environment

The used tool for the ML algorithms implementation and evaluation is the Weka tool [17]; an open source tool dedicated for machine learning algorithms implementation and evaluation. When evaluating the forest fires prediction and detection systems, researchers consider various datasets. The mostly used ones are the satellite data, the meteorological data (collected by local sensors) and the vision-based data (cameras for infrared and smoke captures). We have considered in this study the meteorological data that can be collected in real-time and with very low cost compared to the satellite and vision-based data. The used dataset is described in the following:

1) Dataset Description

The used dataset is the publicly available Algerian forest fires dataset [16]. The constructed dataset includes the daily meteorological observations and the FWI system components. This dataset regroups the data of the Sidi Bel-Abbas and the Bejaia regions located in the northwest and the northeast of Algeria, respectively. The dataset includes 244 instances (138 'fire' classes and 106 'not fire' classes). For the data construction, we have used the meteorological observations for the period of summer of 2012 (from June to September) since the 2012 was the year where the recorded fire occurrence was the highest from 2007 to 2018. The collected weather observations include the critical weather elements influencing the forest fires occurrence and spread, namely temperature, relative humidity (RH), wind speed and rain. The classification is performed using the Canadian fire danger rating system. The FWI system includes six components, namely the FFMC, the Duff Moisture Code (DMC), the Drought Code (DC), the Initial Spread

Index (ISI), the Buildup Index (BUI) and the Fire Weather Index (FWI). The FWI components are calculated using the Fire Weather Index Calculator [18]. Table 1 shows a summarized description of the constructed dataset features. The dataset is publically available in the UCI repository at: https://archive.ics.uci.edu/ml/machine-learning-databases/00547/.

2) Attributs Selection

We have considered four datasets in the experiment and the performance evaluation of the applied ML algorithms based models for forest fires prediction. The first used dataset includes only the weather conditions (4 attributs, namely temperature, relative humidity, wind speed and rain), the second dataset includes the FWI system components (6 attributs, namely FFMC, DC, DMC, BUI, ISI and FWI), the third dataset comprises 6 attributs, a combination of the weather conditions attributs and two FWI components (the FFMC and the FWI). The fourth dataset includes 10 attributs (4 attributs related to the weather conditions and the six FWI system components (FFMC, DC, DMC, BUI, ISI and FWI)).

3) Performance Metrics

We considered three metrics, namely accuracy, recall and precision for the comparison of the machine learning algorithms applied for the forest fires prediction. The formulas (1), (2) and (3) are used for the accuracy, the recall and the precision calculation, respectively. Note that the accuracy is the most important metrics when comparing the performance of the ML algorithms, since it gives an indication to the percentage of the correctly classified instances.

$$\text{Accuracy} = \frac{\text{TP} + \text{TN}}{\text{TP} + \text{TN} + \text{FN} + \text{FP}} \tag{1}$$

$$\text{Recall} = \frac{\text{TP}}{\text{TP} + \text{FN}} \tag{2}$$

$$\text{Precision} = \frac{\text{TP}}{\text{TP} + \text{FP}} \tag{3}$$

The terms TP, TN, FP and FN denote the True Positive, the True Negative, the False Positive and the False Negative values, respectively. The recall metric indicates the number of correctly predicted positive cases. In the context of fire prediction the recall value indicates the number of correctly predicted fire occurrence positive cases.

Table 1. Dataset description

Attribute name	Description	Values interval Sidi-Bel-Abbes region	Values interval Bejaia region	Values intervals The two regions
Temperature	Temperature in Celsius degrees	24 to 42	22 to 37	22 to 42
RH	Relative humidity in %	21 to 90	45 to 89	21 to 90
Wind speed	Wind speed in km/h	6 to 29	11 to 26	6 to 29
Rain	Outside rain in mm/m2	0 to 16.8	0 to 8.7	0 to 16.8
FFMC	Fine Fuel Moisture Code index	41.1 to 94.3	28.6 to 90.3	28.6 to 94.3
DMC	Duff Moisture Code index	0.9 to 65.9	0.7 to 54.2	1.1 to 65.9
DC	Drought Code (DC) index	7.3 to 177.3	7.4 to 220.4	7 to 220.4
ISI	Initial Spread Index (ISI)	0.1 to 18.5	0.1 to 19	0.1 to 19
BUI	Buildup Index (BUI)	1.4 to 68	3.3 to 67.4	1.1 to 68
FWI	Fire Weather Index (FWI)	0 to 31.1	0 to 30.2	0 to 31.1

3 Results and Analysis

We have employed the k-fold cross-validation technique for the training criteria. The cross validation technique divides the dataset into subsets. The dataset is divided into k-folds; each fold contains approximately the same number of instances. One fold will be considered one time in iteration. This process is repeated until all folds have been considered for test and for training. We have used $k = 10$ in our experiments.

Each applied ML algorithm is evaluated using the previously mentioned datasets, namely, the dataset that includes only the meteorological observations (4 attributes), the dataset that includes the FWI system components (6 attributs), and the third dataset that comprises 6 attributs (a combination of the weather conditions and two FWI components (FFMC and the FWI). Finally, the dataset that includes 10 attributs (the 4 attributs related to the weather conditions and the six FWI system components). We will expose in the following an analysis of the obtained results for the four datasets.

3.1 Performance and Time Evaluation and Comparison

We present in this section the performance evaluation (in terms of accuracy, precision and recall) results analysis and comparison between the applied ML algorithms. The evaluation is done for four datasets. As highlighted previously, the forest fires prediction is based on the meteorological observations (temperature, relative humidity, wind speed and rain) and the FWI system components. The ML models are used to classify and predict the fire occurrence cases ('fire' or 'not fire'). We also evaluated the model build time for each ML algorithm considering the 4 datasets. In addition to the performance and the model build time evaluation we have also identified the most influencing attributs among the 10 attributs.

1) Evaluation Using the First Dataset (4 Attributs Meteorological Observations)
Figure 1 shows the performance comparison between the ML algorithms based models for forest fires prediction when using the first dataset, which includes only weather data. As shown in this figure the NB gives the best result in terms of accuracy with about 97.5% followed by the DT and the bagging with about 87%. The results for the NB in terms of recall and precision are about 79.9%, 81.4%, respectively. The obtained results in terms of recall and precision for the NB and the BayeNet are comparable to the others ML models results. The obtained results for the DT, logistic regression, BayeNet, RF, ANN/MLP and Meta algorithms (Bagging and AdaBoost) are comparable. Values in range of 82.4% to 87.7% are obtained for the recall and values in range of 82.5% to 88.3% for the precision. The accuracy values are in range 82.4% to 87.7%. The SVM and the KNN algorithms give the worst performance results with about 64.3% (SVM) and 66.8% (KNN) for the accuracy and the recall and about 63.9% (SVM) and 66.6% (KNN) for the precision. In the whole the DT and Bagging based models give better performance in terms of recall and precision (around 87%).

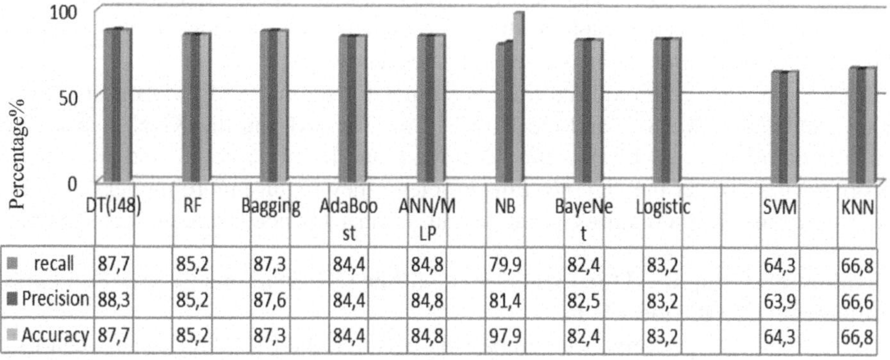

	DT(J48)	RF	Bagging	AdaBoost	ANN/MLP	NB	BayeNet	Logistic	SVM	KNN
recall	87,7	85,2	87,3	84,4	84,8	79,9	82,4	83,2	64,3	66,8
Precision	88,3	85,2	87,6	84,4	84,8	81,4	82,5	83,2	63,9	66,6
Accuracy	87,7	85,2	87,3	84,4	84,8	97,9	82,4	83,2	64,3	66,8

Fig. 1. Performance comparison using the first dataset (meteorological observations)

Regarding the model build time the NB classifier requires less time (with about 0.02 s) compared to the others ML algorithms. It is noticeable that the model build time is comparable for almost all ML algorithms; except for the RF (0.28 s) and the

ANN/MLP (0.2 s); this is due to their complexity. The best time is obtained for the KNN (0 s); however the performance of this ML algorithm is bad when considering the first dataset.

2) Evaluation Using the Second Dataset (6 Attributs: FWI Components)
Figure 2 shows the performance comparison between the ML algorithms when using the second dataset. The results shown in Fig. 2 indicate that the best result is obtained for the ANN/MLP, RF, the LR, and the Meta algorithms with percentage of about 98% for the three metrics. The DT gives results about 96.7% for the tree metrics, the values obtained with the DT are comparable to the best values obtained. The SVM gives the worst results (percentage of 81.1% for the recall and accuracy and percentage of about 85.9% for the precision); however these results are better than those obtained with the first dataset.

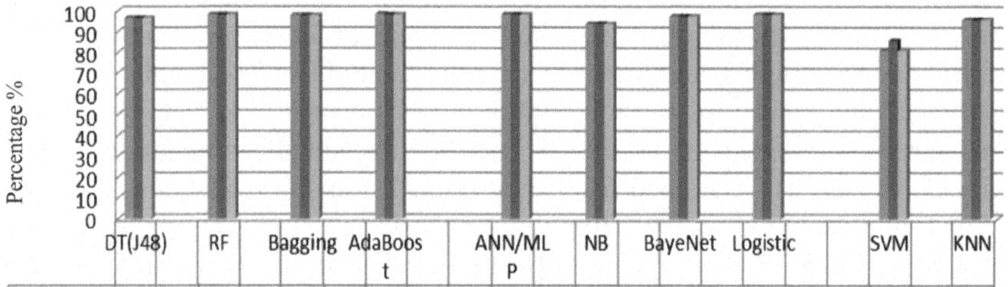

	DT(J48)	RF	Bagging	AdaBoost	ANN/MLP	NB	BayeNet	Logistic	SVM	KNN
recall	96,7	98,4	98	98,6	98,4	93,9	97,5	98,4	81,1	95,9
Precision	96,7	98,4	98	98,4	98,4	94	97,5	98,4	85,9	95,9
Accuracy	96,7	98,4	98	98,4	98,4	93,9	97,5	98,4	81,1	95,9

Fig. 2. Performance comparison using the second dataset (FWI system components)

The NB is not the best classifier when using the second dataset however; it gives good results that are comparable to the other algorithms with about 93.9%, 94%, and 93.6% for the recall, precision and accuracy, respectively.

In terms of time, the required time is similar for the most ML algorithms, except for the ANN/MLP (0.38 s) and the RF (0.13 s). The NB and the KNN give the best results in terms of time with 0 s. The ANN/MLP and RF are accurate and give the best performance, however they are not time efficient, they require more model build time. The bagging algorithm requires less time and its performance is among the best one.

3) Evaluation Using the Third Dataset (Weather Observations (4 Attributs) and 2 Components FFMC and FWI)
Figure 3 shows the performance comparison between the ML algorithms when using the third dataset that includes 6 attributs, a combination of weather observations (4 attributs) and 2 FWI components (FFMC and FWI). When considering the third dataset, the NB and the logistic regression give the best results in terms of accuracy with about 97.4%. The obtained results in terms of recall and precision are the best for the logistic regression, but not the best for the NB. The logistic regression, Bagging, AdaBoost and RF give somehow similar results with percentages around 97% for the three metrics,

followed by the DT with a value of 96.7% for the three metrics, and the BayeNet with values about 96.3% for the three metrics. The ANN/MLP obtained results are about 95.9% for the accuracy and the precision and about 95.6% for the recall. Considering the third dataset, the SVM also give the worst results with 70.9%, 77.9% and 70.9% for recall, precision and accuracy, respectively. Regarding the model build time, the ANN/MLP (0.31 s) requires the highest time followed by the RF (0.15 s) and the SVM (0.1 s). The DT model build time is 0.04 s, even though the model build time is quite similar, the required time for the NB (with 0.02 s) is better than the other models. The NB is time efficient also for the third dataset.

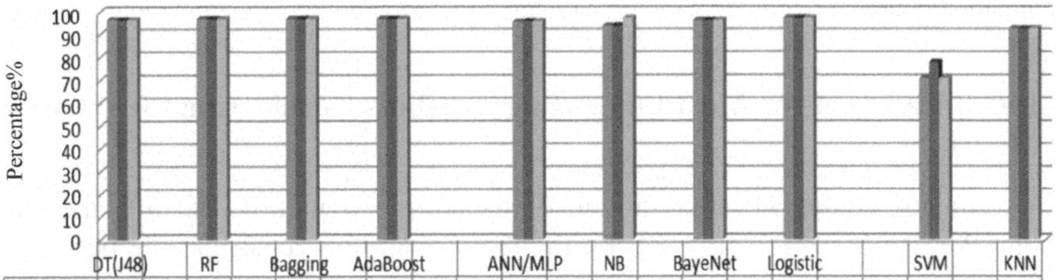

	DT(J48)	RF	Bagging	AdaBoost	ANN/MLP	NB	BayeNet	Logistic	SVM	KNN
recall	96,7	97,1	97,1	97,1	95,6	93,8	96,3	97,5	70,9	92,6
Precision	96,7	97,1	97,1	97,1	95,9	93,9	96,3	97,5	77,9	92,6
Accuracy	96,72	97,13	97,13	97,13	95,9	97,45	96,31	97,54	70,9	92,63

Fig. 3. Performance comparison using the third dataset

4) Evaluation Using the Fourth Dataset (10 Attributs: Weather Observations (4 Attributs) and FWI System Components (6 Attributs))

Considering the fourth dataset, the RF gives the best results (98.4%) that are similar to the results obtained with the Bagging and the AdaBoost, followed by the ANN/MLP with (97.1%), the BayeNet (96.7%), the DT and logistic regression (95.9%). The KNN also gives acceptable results that are comparable to the results obtained with the NB. The NB classifier gives results about 93% for the three metrics. The SVM model gives the worst results with 67.6%, 79.4% and 67.6% for recall, precision and accuracy, respectively.

Regarding the model build time the Bagging (0.5 s), the ANN/MLP (0.37 s) and the RF (0.17 s) require the highest times. The model build time for the DT model is 0.04 s. The smallest time is obtained for the KNN and the NB (0.01 s). Even though the RF and bagging give better performance, their model build time is higher compared to the other ML algorithms (Fig. 4).

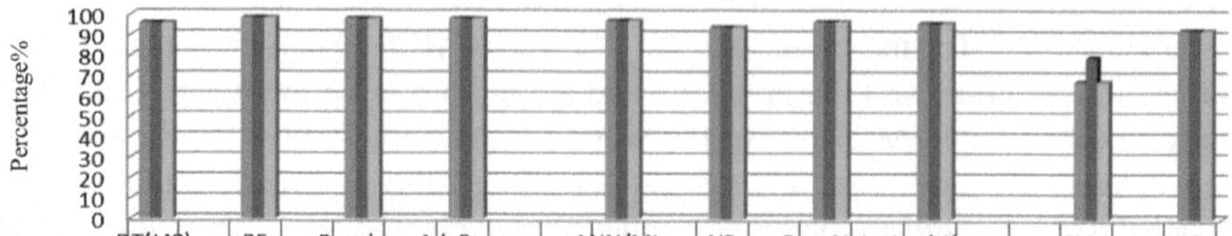

	DT(J48)	RF	Bagging	AdaBoost	ANN/MLP	NB	BayeNet	Logistic	SVM	KNN
recall	95,9	98,4	98	98	97,1	93,9	96,7	95,9	67,6	92,6
Precision	95,9	98,4	98	98	97,2	93,8	96,7	95,9	79,4	92,6
Accuracy	95,9	98,4	98	98	97,1	93,9	96,7	95,9	67,6	92,6

Fig. 4. Performance comparison using the fourth dataset (weather conditions (4 attributs) + FWI components (6 attributs))

3.2 Identification of the Most Influencing Attributs Considering the DT Based Model

We have considered the DT based model, as an example, to identify the most influencing attributs. We summarize in this section the main outcomes regarding the performance results (in terms of recall, precision and accuracy).

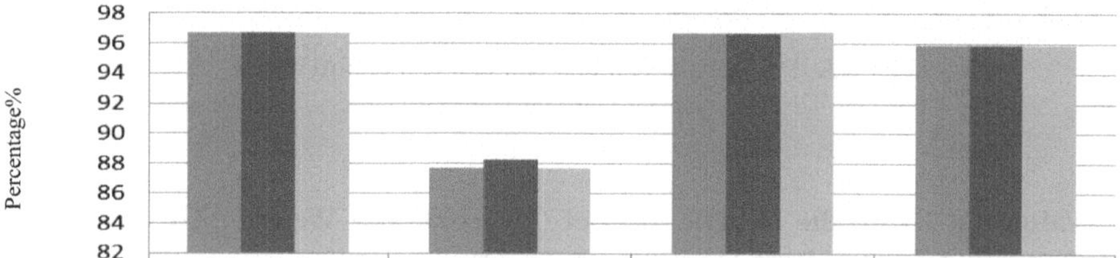

	DT_6_fwi_c	DT_4_att_m	DT_6_at_f_m	DT_10_at
Recall	96,7	87,7	96,7	95,9
Precision	96,7	88,3	96,7	95,9
accuracy	96,7	87,7	96,72	95,9

Fig. 5. Attributs influencing the results considering DT based model

The results summarized in Fig. 5, show that the best performance considering the DT based model is obtained when using the third dataset, which includes 6 attributs (a combination of the weather observations (4 attributs) and two FWI components (FWI and FFMC)). The results are similar to those obtained for the second dataset (6 attributs: the FWI system components only). In fact the information related to the meteorological observations is included in the FWI system components. The results when using the first (4 attributs/weather conditions) and the fourth dataset (10 attributs) are less than those obtained with the second and third dataset. Since the third dataset includes only two FWI components (FFMC and FWI), we can conclude that the other FWI components, namely DMC, DC, ISI and BUI are irrelevant attributs, i.e., they did not influence in the performance evaluation and the prediction results.

Note that compared to the evaluation using the others datasets, the poorest results are obtained with the first dataset (4 attributs, only weather observations). However, these results are quite comparable (around 87% for the three metrics). The most influencing attributs are the FWI, the FFMC and the weather attributs (temperature, relative humidity, wind speed and rain) when combined with the two FWI system components, namely the FWI and the FFMC.

4 Conclusions

In this paper, we have exposed an analytic and comparative study between several machine learning algorithms based models for forest fires prediction. We have identified the most efficient ML algorithms for forest fires prediction in Algeria and the most influencing attributs, i.e. the relevant attributs considering several datasets. We summarize in the following the most important outcomes of this study.

The performances of almost all applied ML algorithms as forest fires prediction models is above 80% for the first dataset (except the SVM algorithm), and above 92% for the three other datasets, thus the ML algorithms based models are reliable and suitable for the forest fires prediction purpose.

This study revealed that the naive Bayesian and the decision tree are suitable for the forest fires prediction purpose, since they give significant performance in terms of accuracy, recall and precision and allow tradeoff between performance and model build time. The obtained results revealed that, the NB is the most time efficient ML based model and gives good performance in terms of accuracy, mainly when weather conditions only were considered (the first dataset). The probabilistic aspect of the Bayesian classifier makes it adaptable to the unpredictable nonlinearity and variability of the data related to natural phenomena such as weather conditions.

The ANN/MLP and the Meta ML algorithms, namely the RF, the Bagging and the Boosting give good performance, which are somewhat better than the performance obtained with the NB and the DT, however they are not time efficient. In the whole the performance results for almost all algorithms are good when considering the second dataset that includes the FWI components (6 attributs).

The obtained results with the second and the third datasets are similar. These results are better than those obtained with first dataset. The fourth dataset that includes 10 attributs gives comparable results, however they are less than the obtained results with the second and the third datasets. When considering the FWI system components (second dataset) the ANN/MLP and RF models perform better; however they are not time efficient. In this case also the NB gives good performance, which is comparable to the best results. The worst results are obtained with the SVM algorithm for all used datasets, except for the second dataset.

Regarding the influencing attributs, this study revealed that the most influencing attributs are the FWI, FFMC components and the weather attributs (temperature, relative humidity, wind speed and rain) when combined with these two FWI system components.

As future work we aim to extend the dataset in terms of size and attributs types for the performance evaluation of the ML algorithms based models. In addition to the weather observations and the FWI components, we aim to include other attributs such as the anthropological factors (human factors).

References

1. Abid, F.: A survey of machine learning algorithms based forest fires prediction and detection systems. Fire Technol. **57**(2), 559–590 (2021). https://doi.org/10.1007/s10694-020-01056-z
2. Maksimović, M., Vujović, V.: Comparative analysis of data mining techniques applied to wireless sensor network data for fire detection. J. Inf. Technol. Appl. **3**(2), 65–773 (2013)
3. Yongqi, P., Yudong, L., Zhongke, F., Zemin, F., Ziyu, Z., Shilin, C., Hanyue, Z.: Forest fire occurrence prediction in china based on machine learning methods. Remote Sens. **14**(21), 5546 (2022). https://doi.org/10.3390/rs14215546
4. Saoudi, M., Bounceur, A., Euler, R., Kechadi, T.: Data mining techniques applied to wireless sensor networks for early forest fire detection. In: Proceedings of the International Conference on Internet of things and Cloud Computing, pp. 1–7 (2016). https://doi.org/10.1145/2896387.2900323
5. Hefeeda, M., Bagheri, M.: Wireless sensor networks for early detection of forest. In: Proceedings of the IEEE International Conference on Mobile Ad Hoc and Sensor Systems, pp. 1–6 (2007)
6. Liu, Y., Gu, Y., Chen, G., Ji, Y., Li, J.: A novel accurate forest fire detection system using wireless sensor networks. In: Proceedings of the IEEE 7th International Conference on Mobile ad-hoc and Sensor Networks, pp. 52–59 (2011)
7. Stojanova, D., Panov, P., Kobler, A., Džeroski, S., Taškova, K.: Learning to predict forest fires with different data mining techniques. In: Proceedings of the 9th International Multi Conference Information Society (2006)
8. Cortez, P., Morais, A.: Data mining approach to predict forest fires using meteorological data. In: Neves, J., Santos, M.F., Machado, J. (eds.) New Trends in Artificial Intelligence. 13th Portuguese Conference on Artificial Intelligence, pp. 512–523 (2007)
9. Abid, F., Izeboudjen, N.: Predicting forest fire in Algeria using data mining techniques: case study of the decision tree algorithm. In: Ezziyyani, M. (ed.) Advanced Intelligent Systems for Sustainable Development (AI2SD'2019). AI2SD 2019. Advances in Intelligent Systems and Computing, Springer, Cham, vol. 1105, pp. 363–370 (2020)
10. Benkheira, A.: Forest fires in Algeria analysis and perspectives, Technical Report, Ministry of Agriculture and Rural Development and Fisheries, Forest Department, Algeria (2018)
11. Dreyfus, G.: Neural networks methodology and applications. Springer, Heidelberg (2005)
12. Kleinbaum, D.G., Klein, M.: Logistic regression: a self-learning text, pp. 1–513, 2nd edn. Springer (2010)
13. Quinlan, J.R.: Induction of decision trees. Mach. Learn. **1**(1), 81–106 (1986)
14. Breiman, L.: Mach. Learn. **45**(1), 5–32 (2001). https://doi.org/10.1023/A:1010933404324
15. Vapnik, V.N.: An overview of statistical learning theory. IEEE Trans. Neural Netw. **10**(5), 988–999 (1999)
16. https://archive.ics.uci.edu/ml/machine-learning-databases/00547/
17. Weka Tool. https://www.cs.waikato.ac.nz/ml/weka/
18. Fire Weather Index Calculator: https://fire.synopticlabs.org/tools/fwi/

The Benefits of Gamification and Mobile Learning in Programming Complex Data Structures in the Context of the Coronavirus Disease

Samia Drissi[1,2(✉)], Karima Boussaha[3], and Aida Chefrour[1,4]

[1] Mohamed Cherif Messaadia University, Souk Ahras, Algeria
{s.drici,aida.chefrour}@univ-soukahras.dz
[2] LiM Laboratory, University of Souk Ahras, P.O. Box 1553, 41000 Souk-Ahras, Algeria
[3] University of Oum El Bouaghi, Oum El Bouaghi, Algeria
karima.boussaha@univ-oeb.dz
[4] LISCO Laboratory, Badji Mokhtar University, B.P-12, 23000 Annaba, Algeria

Abstract. Today, mobile technology is playing an integral part of individuals' lives especially in education. To satisfy the learning needs of learners, active pedagogies is the main objective of mobile learning. From this perpective, to promote their motivation through active pedagogies in educational settings, gamification has garnered significant interest for various information technologies including mobile technology. Therefore, in this research paper, our main contribution brings up is to investigate how incorporating gamification with mobile technologies enhences motivation of learners in higher education, and focuses on how gamification can assist students in assimiling and programing complex data structures lessons during Coronavirus period.

Keywords: Mobile learning · gamification · motivation · computer programming · coronavirus

1 Introduction

The Coronavirus pandemic has caused massive damage around the world, not only to the physical well-being of individuals but also to economies, politics, and society [1]. The application of the lockdown measures has impacted numerous fields like that of higher education, which impacts all university activities including learning and teaching tasks [2]. To ensure the uninterrupted provision of quality education, universities have been migrated to the online mode.

As an example, the Souk Ahras University which located in the east of Algeria had also moved from traditional face-to-face instruction to online modalities especially in computer science field using learning management systems, video conferencing tools like Google Meet. Despite all efforts applied, our department of Computer Science

M. R. Laouar et al. (Eds.): ICISAT 2023, LNNS 981, pp. 111–120, 2024.
https://doi.org/10.1007/978-3-031-60591-8_10

have marked a noticeable increase in dropout rates and a decline in student engagement and motivation particularly in modules that involve problem-solving, computer programming, and algorithmcs.

Nowadays technological advancements particularly mobile technology has brought speed transformations in several aspects of people's lives including the educational field. Incorporating mobile technology in institutions has emerged as a very significant instrument in facilitating the learning and teaching process [3, 4]. Thus, Mobile learning has become a crucial instructional platform in many educational institutions worldwide, including schools, colleges, and universities [5].

Maintaining student motivation and engagement of students in online mode remains a significant challenges for educators around the world encounter [6]. In response to this challenge, There is an extensive literature showing the effects of gamification as a pedagogical appraoch aiming to favor student engagement [7] and foster motivation and participation among students [8]. Moreover, the integration of gamification not aims only to decrease the the negative emotions addressed in conventional educational methods, but it serves in improving students' capacity in acquiring new skills [9] and enhancing problem-solving skills [10] especially in teaching computer science courses [11].

In the higher education context, a growing trend among educators to adopt gamification and mobile learning [7, 12] to promote intrinsic motivation of and optimize learners achievements more specifically in computer science lessons like programming field [13].

Therefore, in this research, we will focuse on the development of a Gamified Mobile Leaning Framework (GAME_ML_FW) serving as an mobile application aiming at teaching and evaluating students in the programmation of data structures courses. GAME_ML_FW is developed as an motivating teaching application during Coronavirus period.

Our primary objective is to investigate the effect of gamification on the motivation of learners in higher education by focusing on the context of learning programming complex data structures. To address this objective, an experiment study was conducted comparing the motivation scores of two groups of students using two versions of GAME_ML_FW. The first one, referred to as the control group, used GAME_ML_FW -V1 without gamification. While the second served as an experimental group using GAME_ML_FW –V2 integrating gamification. Comparing the motivation scale scores between the two groups aims to determine if there is a significant difference in motivation levels in the context of learning programming complex data structures.

In brief, The goal of this empirical study is to give an answer this research question: Is there a significant difference in the motivation scale scores between computer science students who study with GAME_ML_FW –V2, which integrates gamification, compared to the control group styding with GAME_ML_FW -V1 without gamification ?

The present paper is outlined as follows. İn the next section, we will provide some related works on the myriad of mobile learning and gamification during the period of coronavırus pandemic. The description of our Gamified Mobile Leaning Framework (GAME_ML_FW) will be presented in Sects. 3. Section 4 will be dedicated to present the experimental results obtained for the evaluation process. Finally, discussions and conclusions will be drawn.

2 Related Works

2.1 Educational Challenges in the Coronavırus Pandemic: Incorporating Gamification with Mobile Learning Systems as an Example

Mobile Learning (ML) which allows students to access information and knowledge through mobile devices [4], has used as a valuable educational platform, particularly during the COVID-19 pandemic [3]. However, during this critical time, educators are faced with the challenge of developing learning environments that promote active participation [14], enhance motivation [15], and facilitate a transition from memorization-based learning to substantial learning [16].

To address these challenges, gamification which defined in [17] as the integration of game mecanismes to non-gaming contexts has demonstrated its efficacy in fostering motivation and bolstering active participation among university learners, especially in the confinement period [15]. Hence, in the light of the coronavırus issue, incorporating gamification with mobile learning systems has garnered significant interest and has been adopted for learning several fields including chemistry [18], Biology [19]and computer science [20].

2.2 Gamification and Its Contribution to Facilitating Computer Programming Difficulties for Students

Without any doubt, Computer science algorithms can be challenging for learners, and traditional teaching methods may not always be effective in facilitating understanding. İndeed, the use of gamification in computer science descipline is one of the impotant pedagogical approaches which aims to streamline complex topics and foster students' involvement.

Computer programming is a vital skill in computer science which helps students understand abstract concepts by allowing them to apply theoretical knowledge in practical scenarios [21]. Moreover, computer programming teaches problem-solving, logical reasoning, and critical thinking [22]. These competencies make students feel difficult during their studying and lead to frustration and demotivation. These negative feelings may contribute in increasing dropout and avoidance rates among students.

To reduce the negative emmotions and improve overall retention and success rates of students, educators mention that gamification is very effective in subjects requiring a high level of engagement like programming courses [23]. Towards this direction, the study conducted by [24] shown that gamification is a good ways to motivate students in learning programming courses. They assessed how gamification affected the learning process by comparing two groups(a non-gamified and a gamified groups). They found a significant improvements in terms of participation and motivation.

Moreover, [25] explored the importance of applying gamification in computer science by implementing an application Code Combat as an engaging tool for learning object-oriented programming using various gaming elements. Additionally, they found that their application plays an effective role in engaging students and making the progamming process very interesting.

Furthermore, [13] developed an application that aims to evaluate the proficiency of trainees in the C programming language. İn their work, they integrates badges and provides a customized recommendation with gamers. Moreover, the effectiveness and the utility of this application were assessed by students who highlighteda s beneficial influence on the learning process and an usefulness in higher education for enhancing their understanding.

2.3 How Does Gamification İmprove Motivation?

Prensky [26] asserts that motivation is considered as a crucial factor of a successful learning process, but many students consider traditional school education uninteresting and boring. In this regard, numerous researches have shown that the incorporation of game mechanisms has a positive impact on learners motivation [25, 13]. On the other hand, it should be noted that not all gamified frameworks have demonstrated positive effects on motivation. As highlighted in [27], it was determined that the use of badges did not yield significant effects on student motivation.

To conclude, It is clair from literature review section that various research studies have highlighted divergent perspectives on the effectiveness of gamification, which indicates that there is a lack of understanding about factors that contribute to increased effectiveness of gamification in learning contexts, particularly in programming data structures lessons. To this end, in this study, the framework GAME_ML_FW will be presented. GAME_ML_FW adopts motivational strategies for promoting active learning and solving motivational problems by employing a gamification module using different gamified elements like: scores, levels, badges, feedback and leader board.

3 Materials and Methods

3.1 Context and Materials

Our aim is to present a Gamified mobile learning framework named GAME_ML_FW. The goal of this latter is to motivate students in programming various data structures in Java language such as: Stacks, Queues, Hash table, Linked list, Binary tree, Search tree, TAS tree, Balanced Tree, and graph.

GAME_ML_FW offers learners two phases: theoretical phase that focuses on theoretical concepts, and problem-solving phase, aiming to consolidate theoretical concepts by practicing each data structure.

With GAME_ML_FW, as shown in Fig. 1, students can select a data structure starting with linked lists. The student watches a video that offers definitions and examples of linked lists. Next, they investigate Linked List operations such as addition, deletion, and element search, identifying the matching method for each action. This theoretical goal aids students in comprehending data structures and the uses for which they are intended.

In the same phase, the self-assessment of the learner is conducted by answering questions in a theoretical test related to a specific data structure. Then, the learner's answers are automatically evaluated, and a theoretical score (TH_SC) is given for each theoretical test based on a maximum possible score of 10 points.

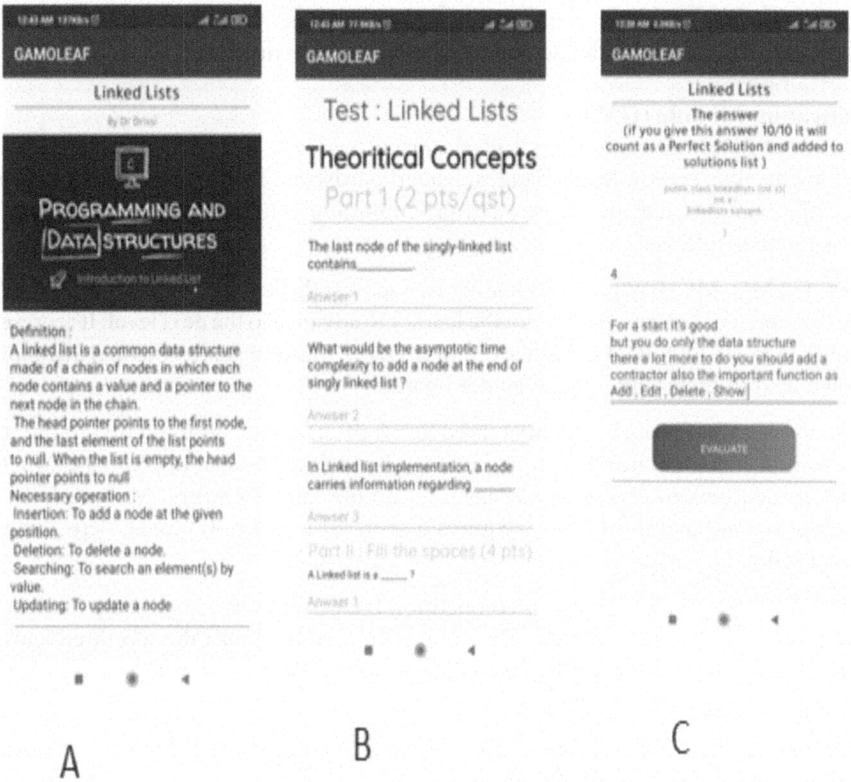

Fig. 1. Some screenshots from GAME_ML_FW

In the second phase, the learner passes to paratice a problem relating to Linked Lists by writing a code under the Java language to solve the proposed problem. Taking an example on linked lists. Student should begin by declaring a class describing the data structure, declaring its attributes, any other necessary variables and its constructors to initialize its object. Then, the student shoud implement its methods by writing the code for each method such as: inserting a node, deleting a node, searching for a value, displaying the Linked List, etc. And finally, student can test each method by calling them from the main class.

During the evaluation process, when a learner submits their solution the teacher is notified. He evaluates the the solution and then a problem-solving score (PS_SC) calculated out of 10 points and assigned to the learner's solution. If the teacher finds a solution to be correct and perfect, he can add it to the best olutions list serving as a repository of solutions that learners can access. İn contrast, if the teachers find that the given solution has shortcomings, a feedback sent to the learners.

İn the learning process, the performance of students is monitored through theoretical tests and problem-solving activities for each data structure they have learned. Therefore, a data structure score is calculated for each data structure by summing the theoretical and the problem-solving scores. This total score can be accumulated and used to unlock subsequent levels. To further increase student engagement, a gamification module (GM) is integrated. This later will be detailed in the next subsection.

Gamification Module (GM)
GAME_ML_FW is a teaching tool used during covid 19 that incorporates gamification elements like, scores, levels,badges, leaderboards, and feedback to engage students in learning complex data structures and programming. These integrated elements are implemented as follows:

Scores GAME_ML_FW provides learner two scores: the level score and the total score. The level score is used to permit or not the learner to advance to the next level. İt represents a number of points obtained in each level. In contrast, the total score obtained in all levels is shown next to their name on the leader board.

Levels the learning process contains three levels of difficulty from the first to the third level, to make our environment challenging. Each level has multiple data structure courses for learners to choose from, ensuring no boredom or worry. Moving from the first to the second and third levels requires a score of at least 30 points, depending on the level score.

Badges the framework employs a competition badge and challenge badge badges by passing from one level to another. The competition badge ranks the top three learners with the highest total score while the challenge badge, based on level score, consists of three medals bronze, silver, and gold.

Leader Board The leader board is used to show the status of all learners' according their obtained level and their total scores.

Feedback a teacher in GAME_ML_FW framework is responsible for assessing a problem-solving solution after notifying the learner of completion. They check attributes, constructors, and methods, awarding a score of 10 and providing feedback on errors and their types.

3.2 Experimental Procedures

To verify our contributions, an experiment is done at Souk Ahras University (Algeria), precisely at the Department of Computer Science. This study involved 60 participants. Both experimental group (ExpG) and the control group (CG) contains thirty (30) participants for each one.

The study utilized Motivated Strategies for Learning Questionnaire (MSLQ) [29], which incldes 8 items rated on a 5-point Likert scale ranged from "1" to "5".

Before conducting the learning activity as a first stage, all students received basic instruction. Next, they were assigned to examine a pre-test motivation questionnaire before beginning instruction. The study involved students in the experimental group using GAME_ML_FW -V2 gamification for three months, while the control group was exposed to GAME_ML_FW -V1. A motivation post-test questionnaire was administered to assess students' motivation after the instruction, starting from October 1st in the 2020/2021 academic year.

4 Results

In this section, the scores the motivation questionnaire will be analyzed using an independent t-test.

Table 1. Group statistics.

	Group	N	M	SD	Std. Error Mean
Pre-test	CG	30	2.87	0.97	0.18
	ExpG	30	2.70	1.09	0.20
Post-test	CG	30	2.60	0.50	0.09
	ExpG	30	3.80	0.61	0.11

N: number of participants, M: mean, SD: standard variations.

Based on the finding depicted in Table 1, it can be observed that during the pre-test phase, the motivation scores of the control group (M = 2.87) and the experimental group (M = 2.70) exhibited proximity to each other. Additionally, the standard deviations for both groups, control (SD = 0.97) and experimental (SD = 1.09), were found to be similar.

Further, it is noted that in the post-test phase, the mean motivation score for the experimental group (M = 3.80) exceeded that of the control group (M = 2.60). Based on the findings, we concluded that there is no significant difference in the motivation pre-test scores between the two groups. This finding is justified by the fact that the Sig (0.113) exceeds the predetermined significance level of 0.05. So, both groups had the identic level of motivation before engaging in the learning process (Table 2).

Table 2. Independent samples test on the motivation pre-test questionnaire.

		Levene's test for equality of variance		t-test for equality of means			Mean Difference & Std. Error Difference
		F	Sig	t	df	Sig. (2-tailed)	
Pre-test	Equal variance assumed	2.588	0.113	0.626	58	0.534	0.167 & 0.266
	Equal variance not assumed	0.534		0.626	57.299		0.266 & 0.266

Table 3 reports the independent samples t-test fidings of motivation post-test questionnaire answered by the two groups. The mean value, standard deviation and standard error of the post-test scores were 3.80, 0.61 and 0.11 for the experimental group and 2.60, 0.50 and 0.09 for the control group. Based on the information provided in Table 3, there was a significant difference (t(58) = − 8.343 and sig. = 0.000) in the scores of the experimental group and the control group.

Table 3. Independent samples test on the motivation post-test questionnaire.

		Levene's test for equality of variance		t-test for equality of means			Mean Difference & Std. Error Difference
		F	Sig	t	df	Sig. (2-tailed)	
Pre-test	Equal variance assumed	0.000	1	− 3.343	58	0.000	− 1.20 & 0.14
	Equal variance not assumed	0.000		0.626	55.769		− 1.20 & 0.14

The results of our study suggest that the incorporation of gamification had impacted the motivation levels of computer science students, specifically in the context of programming complex data structures during the coronavirus period. So, the use of gamification in teaching these concepts increases motivation among the students.

5 Conclusion and Discussion

This study indicates that the gamification approach using scoring, leader board, unlocking levels and feedback resulted in a notable enhancement in students' motivation. Our findings align with other research studies that have shown that the integration of gamification leads to a significant boost in students motivation [6, 8, 27].

References

1. Arnove, R.F.: Imagining what education can be post-Covid-19. Prospects **49**, 43–46 (2020)
2. Cai, P., et al.: The outcomes of lockdown in the higher education sector during the COVID-19 pandemic. PLoS ONE **18**(4), e0282907 (2023)
3. Eom, S.: The effects of the use of mobile devices on the E-learning process and perceived learning outcomes in university online education. E-learn. Digital Media **20**(1), 80–101 (2023)
4. Adzifome, N.S., Agyei, D.D.: Learning with mobile devices-insights from a university setting in Ghana. Educ. Inf. Technol. **28**(3), 3381–3399 (2023)
5. Voicu, M.C., Muntean, M.: Factors that influence mobile learning among university students in Romania. Electronics **12**(4), 938 (2023)
6. Al Breiki, M., Yahaya, W.A.J.W.: Using gamifcation to promote students' engagement while teaching online during COVID-19. In: Fayed, I., Cummings, J. (eds.) Teaching in the post COVID19 era. Springer, Heidelberg (2021)

7. Pardim, V.I., Contreras Pinochet, L.H., Viana, A.B.N., Souza, C.A.D.: Where is the student who was here? Gamification as a strategy to engage students. Int. J. Inf. Learn. Technol. **40**(2), 177–192 (2023)
8. Camacho-Sánchez, R., Manzano-León, A., Rodríguez-Ferrer, J.M., Serna, J., Lavega-Burgués, P.: Game-based learning and gamification in physical education: a systematic review. Educ. Sci. **13**(2), 183 (2023)
9. Jayalath, J., Esichaikul, V.: Gamification to enhance motivation and engagement in blended eLearning for technical and vocational education and training. Technol. Knowl. Learn. **27**(1), 91–118 (2022)
10. Asigigan, S.İ, Samur, Y.: The effect of gamified STEM practices on students' intrinsic motivation, critical thinking disposition levels, and perception of problem-solving skills. Int. J. Educ. Math. Sci. Technol. **9**(2), 332–352 (2021)
11. Jawad, H., Tout, S.: Gamifying computer science education for z generation. Information **12**, 453 (2021)
12. Al Amri, M.M., Almaiah, M.A.: The use of mobile gamification technology for sustainability learning in Saudi higher education. Int. J. **9**, 8236–8244 (2020)
13. Troussas, C., Krouska, A., Sgouropoulou, C.: Collaboration and fuzzy-modeled personalization for mobile game-based learning in higher education. Comput. Educ. **144**, 103698 (2020)
14. Ahshan, R.: A framework of implementing strategies for active student engagement in remote/online teaching and learning during the COVID-19 pandemic. Educ. Sci. **11**(9), 483 (2021)
15. Hosseini, C., Humlung, O., Fagerstrøm, A., Haddara, M.: An experimental study on the effects of gamification on task performance. Proc. Comput. Sci. **196**, 999–1006 (2022)
16. Şahin, G.: Designing a digital escape room game: an experience of a digital learning tool in basic education. J. Educ. Technol. Online Learn. **6**(4), 925–946 (2023)
17. Song, S.: Gamification and User Engagement on Digital Platforms. McGill University, Canada (2022)
18. Lackey, S.: The Effects of Game-Based Learning on the Motivation and Engagment of Career Technical Education Students in the High School Chemistry Classroom (Doctoral dissertation, University of South Carolina) (2022)
19. Frøland, T.H., Heldal, I., Sjøholt, G., Ersvær, E.: Games on mobiles via web or virtual reality technologies: How to support learning for biomedical laboratory science education. Information **11**(4), 195 (2020)
20. Kalogiannakis, M., Papadakis, S., Zourmpakis, A.I.: Gamification in science education. A systematic review of the literature. Educ. Sci. **11**(1), 22 (2021)
21. Saritepeci, M.: Developing computational thinking skills of high school students: design-based learning activities and programming tasks. Asia Pac. Educ. Res. **29**(1), 35–54 (2020)
22. Miller, D.: The best practice of teach computer science students to use paper prototyping. Int. J. Technol. Innovat. Manage (IJTIM) **1**(2), 42–63 (2021)
23. Zhan, Z., He, L., Tong, Y., Liang, X., Guo, S., Lan, X.: The effectiveness of gamification in programming education: evidence from a meta-analysis. Comput. Educ.: Artif. Int. **3**, 100096 (2022)
24. Figueiredo, J., García-Peñalvo, F. J.: Increasing student motivation in computer programming with gamification. In: 2020 IEEE Global Engineering Education Conference (EDUCON) (pp. 997–1000). IEEE (2020)
25. Karram, O.: The role of computer games in teaching object-oriented programming in high schools-code combat as a game approach. WSEAS Trans. Adv. Eng. Educ. **18**, 37–46 (2021)

26. Prensky, M.: Digital natives digital immigrants. From On the Horizon MCB University Press (9),5. URL: https://goo.gl/XiRaJC (2001). Accessed 05 Feb 2020
27. Kyewski, E., Krämer, N.: To gamify or not to gamify? An experimental field study of the influence of badges on motivation, activity, and performance in an online learning course. Comput. Educ. **118**, 25–37 (2018)

Hybrid Deep Learning Models for Efficient Detection of Depressive Disorders from Twitter Data

Lamia Bendebane[1(✉)], Ikram Boubaker[2], Asma Saighi[3], and Zakaria Laboudi[4]

[1] Research Laboratory on Computer Science ReLa(CS)2, University of Oum El Bouaghi, Oum El Bouaghi, Algeria
lamia.bendebane@univ-oeb.dz

[2] LaSTIC Laboratory, University of Batna, Batna, Algeria
ikrame.boubaker@univ-batna2.dz

[3] LIAOA Laboratory, University of Oum El Bouaghi, Oum El Bouaghi, Algeria

[4] Department of Networks and Telecommunications, University of Oum El Bouaghi, Oum El Bouaghi, Algeria
laboudi.zakaria@univ-oeb.dz

Abstract. Social media occupy an important part of people's lives through which they share their experiences, feelings and interests. This leads to the generation of huge volumes of data, especially in textual form. Several methods for natural language processing tasks have been used so as to extract valuable information from social media. In particular, various deep learning approaches have been proposed to study depression detection through twitter data analysis. In this respect, this study proposes a well-defined methodology for building efficient models that allow predicting depression through Tweets analysis. The idea consists in devising an efficient process for combining different types of neural networks, in order to improve the general performances. Once the training phase performed, we proceed to a deployment phase in which the well-performing variants are used to analyse the sentiments of Twitter users in UK and US during COVID-19 period toward depressive disorders behaviours. The proposal is validated through several experiments and comparisons according to some evaluation metrics. Overall, the obtained results are satisfactory and encouraging.

Keywords: Twitter · Natural Language Processing · Sentiment Analysis · Deep learning · Depressive disorders

1 Introduction

Nowadays, social media occupy an important part of our daily life as many people share their thoughts, feelings and experiences every day, thus the generation of large volumes of data. This makes them an excellent source for conducting researches by analysing social data through several methods, in particular sentiment analysis techniques which were adopted in several fields such as purchases and healthcare.

M. R. Laouar et al. (Eds.): ICISAT 2023, LNNS 981, pp. 121–135, 2024.
https://doi.org/10.1007/978-3-031-60591-8_11

In this paper, we are interested in applying deep-learning techniques for sentiment analysis using Twitter data, in order to detect depressive disorders. The starting point is that many people are ready to share their sentiments and to self-report the symptoms related to their health issues through social media in the hope of finding solutions there. This is why it is believed that social media can help predict mental disorders. Depressive disorder is both a symptom and an illness, characterized by sadness, loss of interest or pleasure, feelings of guilt or low self-worth, disturbed sleep or appetite, feelings of tiredness, and poor ability to concentrate [1]. According to the World Health Organization, depressive disorder is one of the major healthcare issues that are increasing every day. Indeed, millions of people suffer from depression while only a few of them are receiving proper treatment [2]. The contents posted through the Twitter platform reflect the feelings of the corresponding users. Thus, Sentiment Analysis (SA) techniques are usually involved to analyse and understand such feelings. This is a sub-area of the natural language processing (NLP) field which focuses on analysing positive and negative features of text or speech. In this respect, earlier SA-based researches have been conducted by applying classical machine learning algorithms to predict depression through Twitter data such as Random Forest (RF), Support Vector Machines (SVMs), Decision Trees (DT), etc. However, due to the huge increase in data volumes, these techniques have reached their limits in terms of efficacy. That is why deep learning techniques are currently adopted, in particular by involving recurrent neural networks (RNN) variants such as LSTM, GRU, Bi-LSTM and Bi-GRU.

Our study focuses on sentiment analysis of tweets written in English that we collect, with the aim of detecting depression disorders observed in users behaviours. The idea consists in involving simple and hybrid variants of deep learning models to examine their strengths and weaknesses. To achieve these goals, a well-defined methodology is first proposed to build effective models. Then, we move to the deployment stage in which the best performing models are used to analyse tweets in the UK and US during the COVID-19 period regarding depressive behaviours. Our proposal is validated by a set of experiments and comparisons. Overall, the results obtained are very satisfactory, encouraging and promising.

The rest of this paper is organized as follows. Section 2 discusses some related works on the application of deep learning approaches to depression detection using Twitter data. Section 3 presents the general scheme of the proposed methodology for detecting depression using Twitter data. Section 4 summarizes the experimental part and discusses the obtained numerical results. Finally, Sect. 5 provides some concluding remarks and future directions.

2 Related Works

This section reviews some related works that focus on analysing Twitter social data for detecting depressive disorders [3–15]. Almouzini et al. (2019) attempted to detect depressed Arab users using Twitter data [3]. The data were collected by extracting over seven thousand tweets from 97 users in the Gulf region, published between November 18, 2016, and July 23, 2018. The tweets consisted of a mix of Modern Standard Arabic and dialects from the Arabian Gulf. Then, a two-phase manual labelling process was carried out, in order to put real labels on the collected dataset. Next, the collected data are cleaned and pre-processed. After that, the depressive symptoms and online behaviour of Arab users were extracted from the texts of the tweets, in order to distinguish the depressed class from the non-depressed class. For this purpose, four machine-learning algorithms were used: Random Forest, Naïve Bayes, AdaBoostM1, and Liblinear, to determine the binary classification. As a result, the authors have achieved an an Accuracy of about 87.5%. In [4], Shetty et al. (2020) relied on Twitter data to apply different machine learning classifiers (LSTM, CNN, linear SVC, etc.) to the detection of depressive cases and thus decides whether a user's tweet expresses some depression signs or not. As a main result, the best classifier was CNN with that recorded an accuracy of about 95%. In [5], Basiri et al. (2021) have proposed a new hybrid fusion model using five deep classifiers and then combined them to improve the final output using a Meta-learning method to find the polarity of tweets. The followed methodology consisted of several steps. First, the data were collected from two sources: Google Trends and Twitter data (1,056,049 tweets). Then, a fusion model was used by involving DistilBERT, CNN, NBSVM, BiGRU and FastText models. As a result, the proposed model was able to classify tweets as positive or negative with an accuracy of about 85,8%. Kour et al. (2022) [6] have used a hybrid deep learning classifier CNN-BiLSTM to predict depressive cases from Twitter users. As a result, their model could achieve an accuracy of 94.28%. In [7], Carmel et al. (2022) have proposed a new classifier to detect Indian people that likely suffer from depression based on their tweets. In practice, Indians tend to tweet mainly in English, Hindi, or a mixture of these two languages (commonly known as Hinglish). The proposed methodology includes several steps: data collection, pre-procession and analysis through classification models. For this purpose, a Multinomial Naive Bayes algorithm was used as the main classifier and Naive Bayes as the feature vector. In addition, Grid search was involved to perform hyper-parameters optimization. As a result, the proposed model could achieve an accuracy of about 96.15%.

3 The General Scheme of the Proposed Methodology

In order to develop efficient models for depressive disorder detection from Twitter data, we have adopted the following methodology, as represented in Fig. 1.

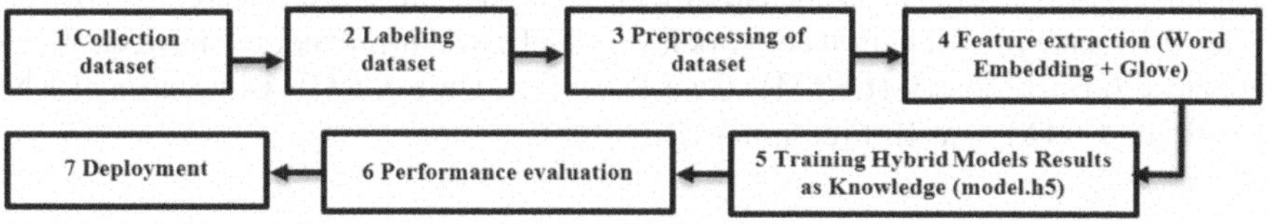

Fig. 1. The proposed methodology for depression detection from Twitter data.

First, we have collected tweets using Twitter API. In total, about 3.596.032 tweets were collected according to different parameters (e.g. keywords, location, language, etc.). These tweets were then subdivided into six sub-datasets namely Train D1, Train D2, Evaluation D1, Evaluation D2, Deployment D1 and Deployment D2, as shown in

Table 1. Some of the keywords used to collect depressive tweets are the following: **"I am depressed, I am diagnosed with depression, I was diagnosed with depression, I have been diagnosed with depression, I am identified as depressed, I was identified as depressed, I have identified as depressed".** For non-depressive tweets, we used other keywords as follows: **"movie, art, food, cinema, sport, animal, politics, society, happiness, joy, optimistic, joke"**. Finally, the keywords used to collect deployment datasets D1 and D2 are: **"covid, covid19, stay at home, corona, coronavirus, I am diagnosed with coronavirus/corona/covid, I was diagnosed with coronavirus/corona/covid, I have been diagnosed with coronavirus/corona/covid"**. The collected tweets were labelled with '0' for the non-depressive class and '1' for the depressive class.

Table 1. The parameters used to collect the sub-dataset and thier labelling.

Datasets	Presented	Tweets	Period	Local	Label
Train D1	Possible depressive	645,159	January 2017 to March 2019	-	1
Train D2	Non-depressive	572,472	September 2019	USA-UK	0
Evaluation D1	Possible depressive	64,410	April to December 2019	-	1
Evaluation D1	Non-depressive	609,725	August 2019	USA-UK	0
Deployment D1	-	437,673	March to May 2020	UK	-
Deployment D2	-	1,266,593	March to May 2020	USA	-

The collected tweets are raw and thus need to be cleaned. For this purpose, we have used several methods to pre-process the labelled data by removing unnecessary contents such as URLs, Hashtags(#), Mentions(@name), special characters, symbols, etc. The resulting datasets are shown in Table 2. Next, we have used different training models: simple and hybrid neural network classifiers. Actually, there are different methods to combine different types of neural network layers such as recurrent neural network (RNN), Long Short Term Memory (LSTM), Gate Recurrent Unit (GRU), Bidirectional RNN, Bidirectional LSTM, and Bidirectional GRU [8, 9].

Table 2. Numbers of tweets before and after pre-processing.

	Train D1	Train D2	Evaluation D1	Evaluation D2	Deployment D1	Deployment D2
Before	645,159	572,472	64,410	609,725	437,673	1,266,593
After	258,580	310,478	19,096	535,750	393,935	1,133,372

In the case of depressive disorder detection, RNNs can be used to process the sequence of words in tweets. Another common approach is to use a convolutional neural network (CNN). CNNs are very suitable for extracting features from image data, but they can also be adapted to extract features from texts, in particular from words in tweets. Finally, it is also possible to rely on a combination of CNN with other models such as RNN, LSTM, GRU, BiRNN, BiLSTM, and BiGRU, to build a hybrid convolutional neural network. The hybrid method is able to extract features from words in tweets as well as to process the sequence of their words. On another side, the Attention layers are a relatively new type of neural network layer that have quickly become popular because of their ability to improve the performance of neural networks for a variety of tasks. Practically speaking, there exist different variants of Attention layers that differ in architectural details. In our study, we used a "Local attention" as it allows dealing with different parts of the input sequences in a local manner [10]. In other words, a local attention layer could be used to deal with different words in a sentence so as to determine the meaning of the sentence, but only within a specific window of words. Although the hybridization of different types of neural network layers may show good performance for predicting depressive disorders from tweets, this is not always guaranteed. Figure 2 illustrates the adopted models used by the proposed method for hybridization namely (1) RNN, (2) LSTM, (3) GRU, (4) BiRNN, (5) BiLSTM, and (6) BiGRU.

The adopted method for hybridization consists in connecting the models shown in Fig. 2 with an attention layer and CNN, as given in Fig. 3.

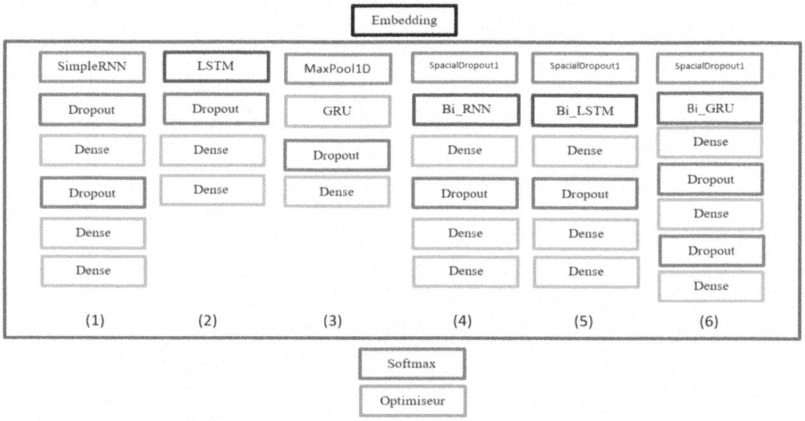

Fig. 2. The models used in our hybrid method.

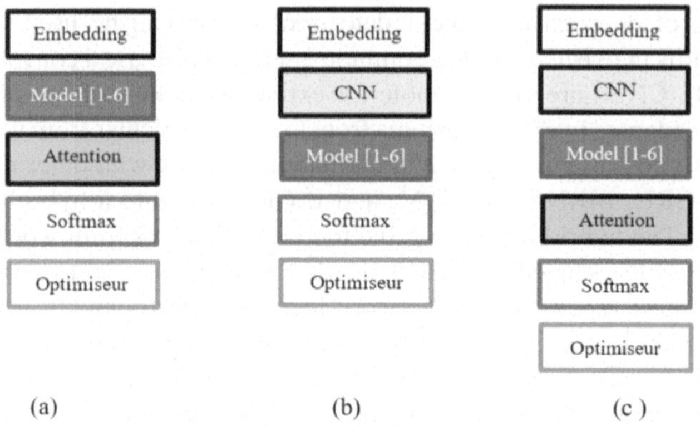

Fig. 3. The adopted method for models hybridization

4 Experiments, Numerical Results, and Discussion

After pre-processing the datasets (tweets) and making them ready for use, they are divided into two subsets: 80% are used to train the models while the remaining 20% are used to validate them. In total, 24 simple and hybrid deep learning models are tested and compared to each other's in terms of Accuracy, Recall, Precision, F1-score, and F2-score metrics. The numerical results are given in Table 3. Moreover, the dimension of the GloVe embedding used is 200 while the hyper-parameters used are 256 of Batch size, 15 epochs, and Adam optimizer with a learning rate of 0.001.

$$Accuracy = \frac{TN + TP}{TN + FP + TP + FN} \tag{1}$$

$$Precision = \frac{TP}{TP + FP} \tag{2}$$

$$Recall = \frac{TP}{TP + FN} \tag{3}$$

$$F_\beta = \left(1 + \beta^2\right) \times \frac{Precision \times Recall}{\left(\beta^2 \times Precision\right) + Recall} \tag{4}$$

	N	**P**
N	TN	FP
P	**FN**	**TP**

N: Negative, P: Positive, T: True, F: False

Table 3. The numerical evaluation of our models using datasets (D1 and D2).

N°	Models	Accuracy	Precision	Recall	F1-score	F2-score
01	SimpleRNN	03.44	03.44	100.00	06.65	15.13
02	**LSTM**	**99.87**	**98.15**	**98.18**	**98.17**	**98.18**
03	GRU	95.30	19.42	11.65	14.56	12.66
04	BiRNN	26.68	04.48	99.95	08.58	19.00
05	**BiLSTM**	**99.84**	**97.73**	**97.69**	**97.71**	**97.70**
06	**BiGRU**	**99.83**	**98.20**	**96.82**	**97.51**	**97.09**
07	RNN_ATT	03.44	03.44	100	06.65	15.12
08	LSTM_ATT	47.30	06.09	99.28	11.47	24.45
09	**GRU_ATT**	**99.20**	**82.42**	**97.64**	**89.39**	**94.16**
10	BiRNN_ATT	80.85	15.15	99.21	26.29	47.03
11	**BiLSTM_ATT**	**99.76**	**96.30**	**96.82**	**96.56**	**96.71**
12	**BiGRU_ATT**	**99.77**	**96.41**	**97.08**	**96.74**	**96.94**
13	CNN_RNN	96.56	00.00	00.00	00.00	NAN
14	CNN_LSTM	91.91	26.10	73.69	38.55	54.00
15	CNN_GRU	97.51	62.73	68.35	65.42	67.15
16	CNN_BiRNN	88.08	22.27	99.02	36.37	58.63
17	**CNN_BiLSTM**	**99.30**	**84.69**	**97.19**	**90.51**	**94.40**
18	**CNN_BiGRU**	**98.56**	**71.44**	**97.09**	**82.31**	**90.58**
19	CNN_RNN_ATT	03.44	03.44	100	06.65	15.12
20	**CNN_LSTM_ATT**	**99.23**	**84.76**	**94.84**	**89.52**	**92.64**
21	CNN_GRU_ATT	97.47	61.53	71.33	66.07	69.13
22	CNN_BiRNN _ATT	97.13	54.56	98.68	70.27	84.94
23	**CNN_BiLSTM_ATT**	**99.22**	**83.05**	**97.27**	**89.60**	**94.05**
24	**CNN_BiGRU_ATT**	**98.86**	**76.16**	**97.28**	**85.43**	**92.17**

According to Table 3, the well-performing models are **LSTM, BiLSTM, BiGRU, CNN**_BiLSTM**, CNN_BiGRU, GRU_ATT, BiLSTM_ATT, BiGRU_ATT, CNN_BiLSTM_ATT, and CNN_BiGRU_ATT** as they record more than **98%** for all performance accuracy metrics. In contrast, the model **SimpleRNN** has the worst performance despite the combination with CNN and the Attention layer. The performance of **GRU** and **BiRNN** models is improved when they are combined with CNN or Attention layer. Figures 4, 5 and 6 illustrate respectively the model training accuracy, model training loss and the confusion Matrix of our models.

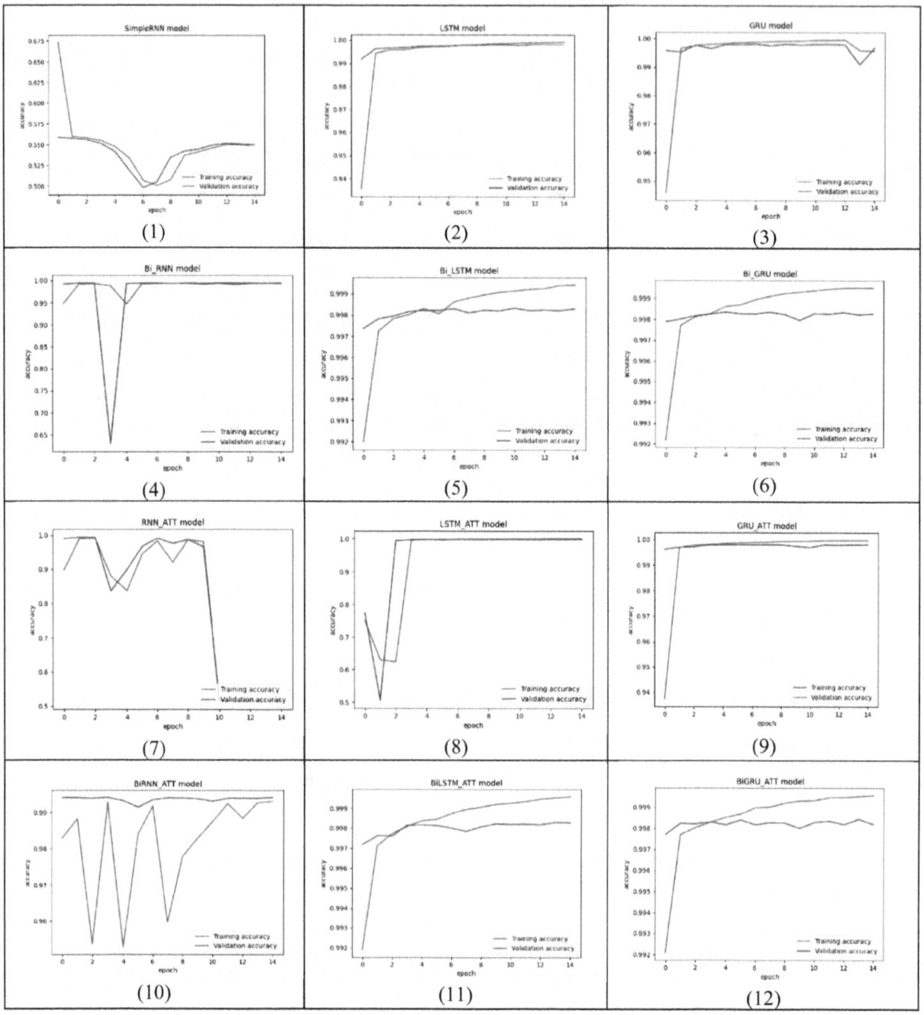

Fig. 4. Model training Accuracy of our models.

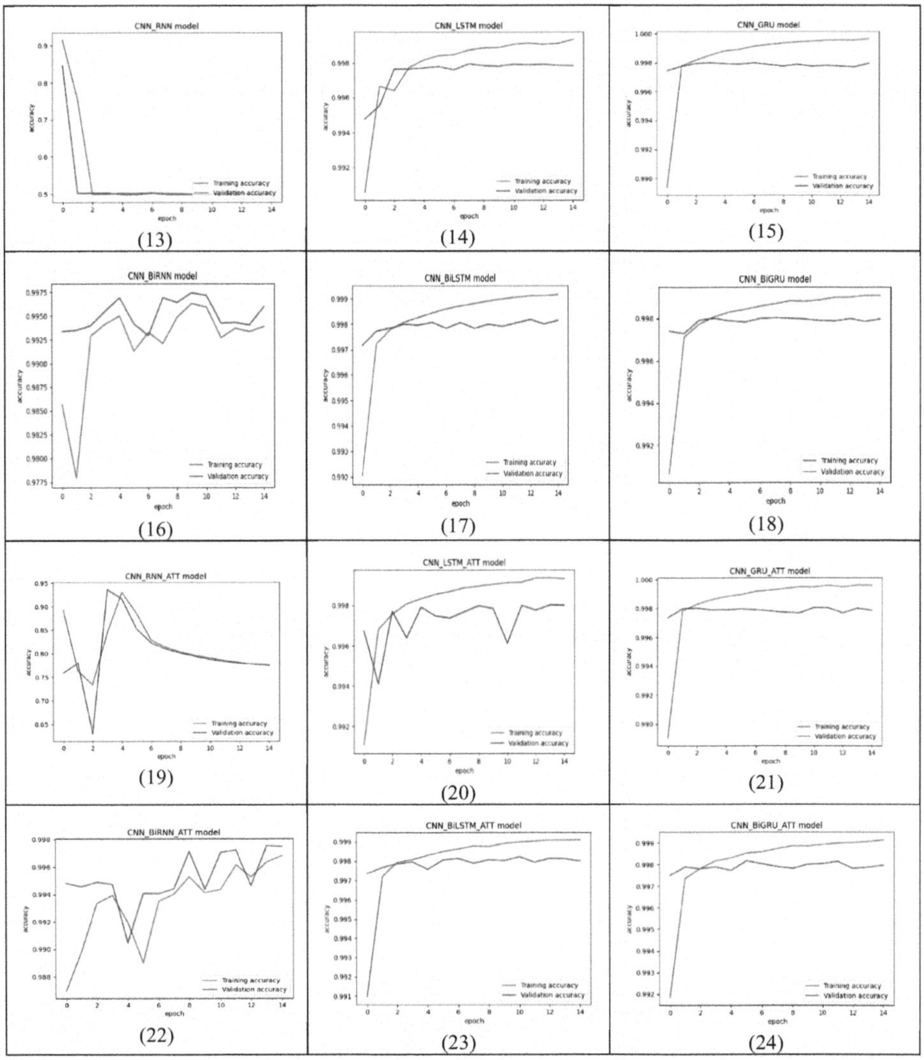

Fig. 4. *(continued)*

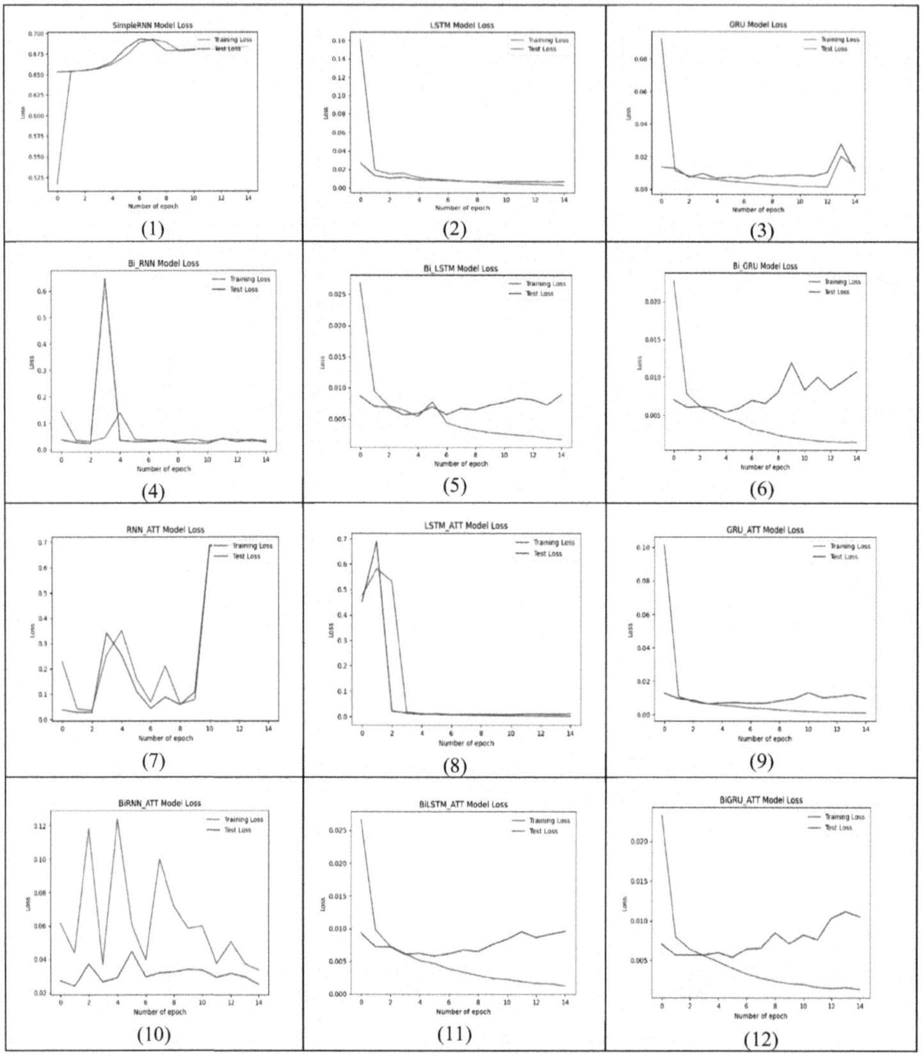

Fig. 5. Model training Loss of our models.

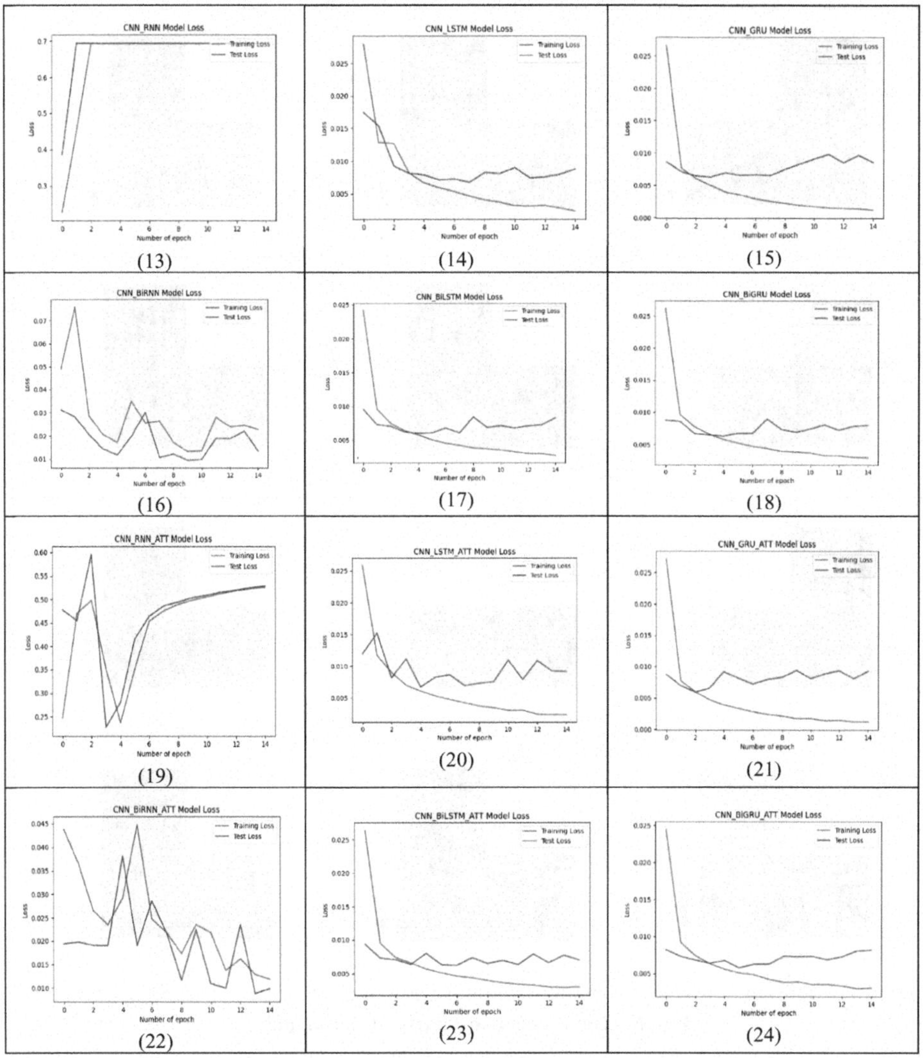

Fig. 5. *(continued)*

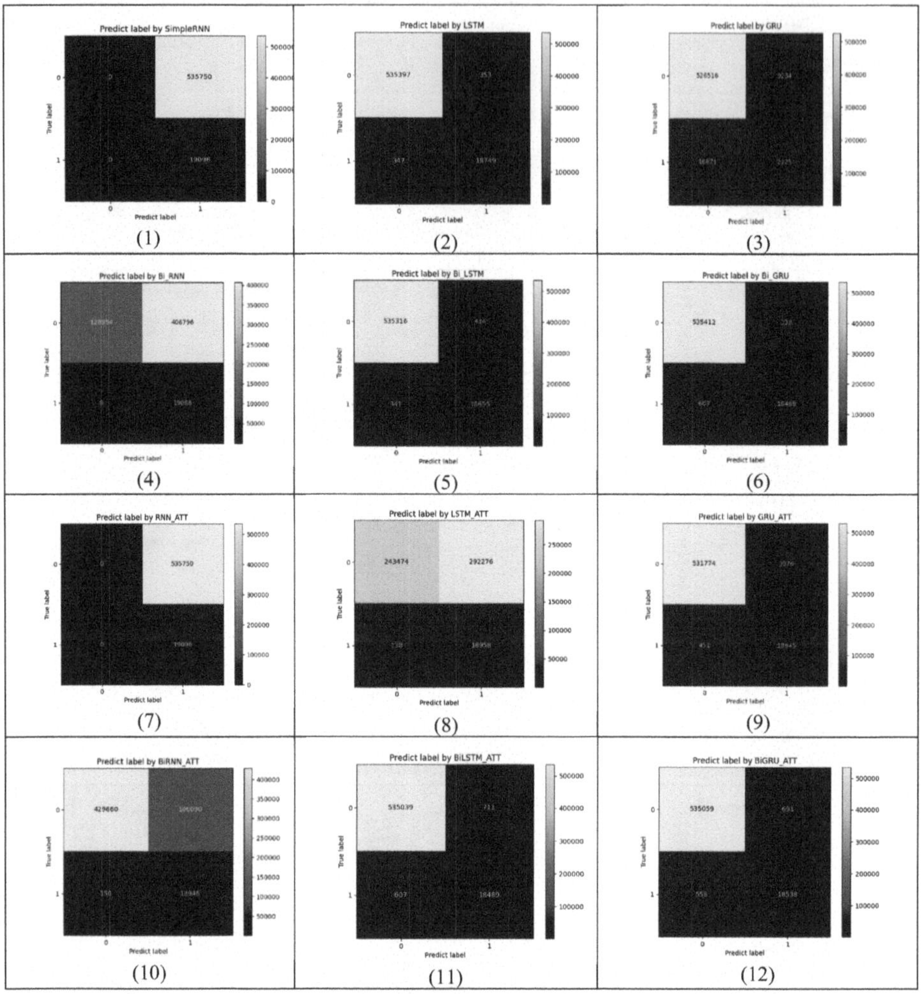

Fig. 6. The Confusion Matrix of our models.

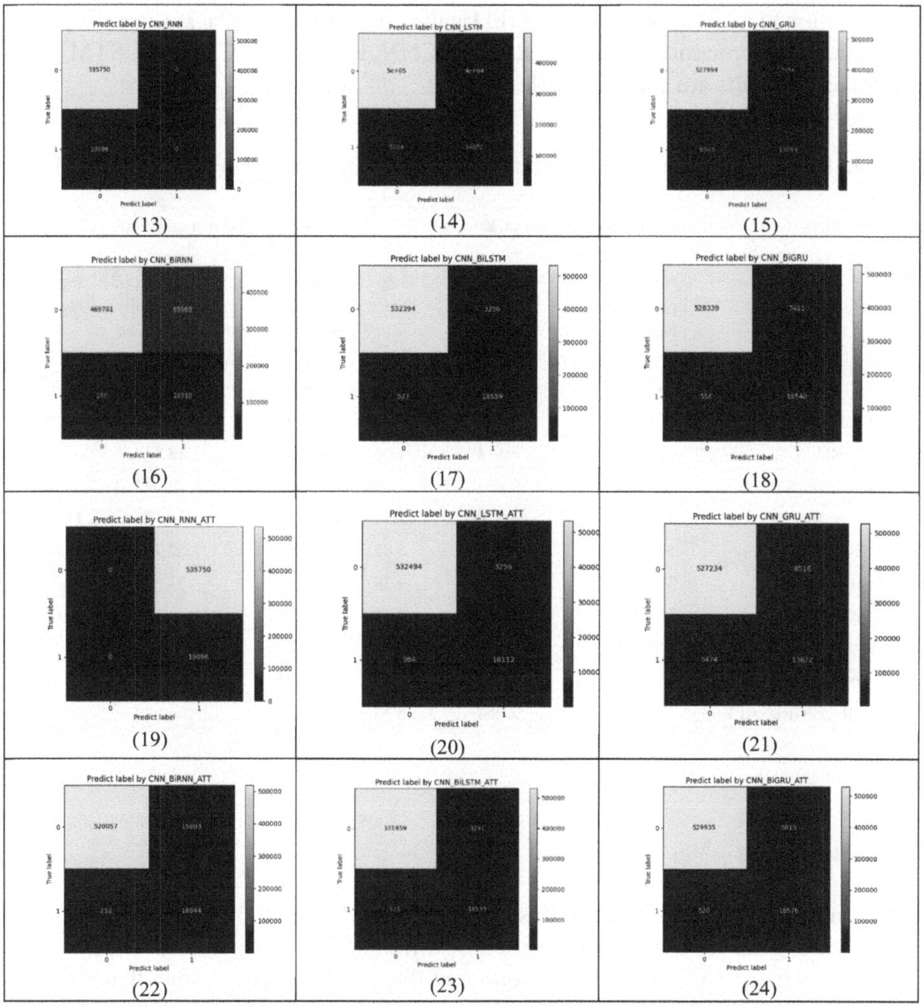

Fig. 6. *(continued)*

Finally, the resulting models of the learning phase are exploited in the deployment phase in which the efficient variant is used to analyse the feelings of Twitter users in the United Kingdom and the United States during the COVID- 19 with regard to depressive behaviour. Indeed, although the deployment phase is a fundamental step in deep learning processes, most related research-works do not consider it. Hence, we have collected a dataset of tweets that were posted during the COVID-19 period in order to try to predict tweets showing depressive behaviours. This choice is motivated by the declaration of the World Health Organization (WHO) according to which the worldwide prevalence of anxiety and depression has increased by 25% during the Covid-19 period. To do so, we use the dataset Deployment D1 from the UK and the dataset Deployment D2 from the USA. In addition, we used the well-performance models to predict the depressive tweets.

As motivation the declaration of the World Health Organization (WHO) 1 in 8 people are diagnosed with depression as predicted by the CNN_BiGRU and CNN_BiLSTM_ATT. The obtained results are given in Table 4.

Table 4. Prediction of depressive tweets during the COVID-19 period in the UK and the USA.

N°	Models	Deployment D1 from UK 393,935 tweets			Deployment D2 from the USA 1,133,372 tweets		
		Number of potentially Non-depressive tweets	Number of potentially depressive tweets	% of Dep	Number of potentially Non-depressive tweets	Number of potentially depressive tweets	% of No-Dep
01	LSTM	392,562	1,373	0.35	1,129,332	4,040	0.36
02	BiLSTM	387,727	6,208	1.58	1,116,941	16,431	1.45
03	BiGRU	387,401	6,534	1.66	1,114,541	18,831	1.66
04	CNN_BiLSTM	364,067	29,868	7.58	1,051,279	82,093	7.24
05	**CNN_BiGRU**	**350,215**	**43,720**	**11.10**	**1,007,594**	**125,778**	**11.10**
06	GRU_ATT	373,255	20,680	5.25	1,080,655	52,717	4.65
07	BiLSTM_ATT	376,687	17,248	4.38	1,081,814	51,558	4.55
08	BiGRU_ATT	378,705	15,230	3.87	1,087,380	45,992	4.06
09	CNN_LSTM_ATT	370,821	23,114	5.87	1,076,725	56,647	5.00
10	**CNN_BiLSTM_ATT**	**354,680**	**39,255**	**09.96**	**1,029,234**	**104,138**	**9.20**
11	CNN_BiGRU_ATT	370,951	22,984	5.83	1,068,683	64,689	5.71

5 Conclusion

The main objective we targeted in this study was the detection of depressive disorders using Twitter data via a new combination of different types of neural network layers. To do so, we have first presented the outline of the adopted methodology that relied on sentiment analysis of Twitter data. Then, we have highlighted the experimental part as well as the obtained results. As a main result, the combination of BiLSTM and BiGRU models with CNN and/or with the local attention layer recorded the best performances. Overall, the results were promising and very satisfactory. As a perspective, we plan to extend our work so that is supports other mental disorders such as anxiety and bipolarity.

References

1. Diagnostic and Statistical Manual of Mental Disorders: Fifth Edition Text Revision DSM-5-TR™. 9780890425770, 9780890425756, 9780890425763, 2021051781, 2021051782. American Psychiatric Association Publishing. (2022)
2. World Health Organization official website. https://www.who.int/news-room/fact-sheets/detail/mental-disorders
3. Almouzini, S., Alageel, A.: Detecting arabic depressed users from twitter data. Proc. Comput. Sci. **163**, 257–265 (2019)

4. Shetty, N.P., Muniyal, B., Anand, A., Kumar, S., Prabhu, S.: Predicting depression using deep learning and ensemble algorithms on raw Twitter data. Int. J. Electr. Comput. Eng. Scopus **10**(4), 3751 (2020)
5. Basiri, M.E., Nemati, S., Abdar, M., Asadi, S., Acharrya, U.R.: A novel fusion-based deep learning model for sentiment analysis of COVID-19 tweets. Knowl.-Based Syst. **228**, 107242 (2021)
6. Kour, H., Gupta, M.K.: An hybrid deep learning approach for depression prediction from user tweets using feature-rich CNN and bi-directional LSTM. Multimed. Tools Appl., Springer **81**(17), 23649–23685 (2022). https://doi.org/10.1007/s11042-022-12648-y
7. Carmel Mary Belinda, M.J., Arif, M.: Linguistic analysis of hindi-english mixed tweets for depression detection. J. Math. **2022**(1), 3225920 (2022)
8. Brownlee, J.: Deep learning with python develop deep learning models on theano and tensorflow using keras. Copyright 2016 Jason Brownlee. All Rights Reserved. Edition: v1.7 (2016)
9. Zeberga, K., Attique, M., Shah, B., Ali, F., Jembre, Y.Z., Chung, T.S.: A novel text mining approach for mental health prediction using Bi-LSTM and BERT model. Comput. Int. Neurosci. **2022**(1), 7893775 (2022)
10. Ghosh, S., Anwar, T.: Depression intensity estimation via social media: a deep learning approach. IEEE Trans. Comput. Soc. Syst. **8**(6), 1465–1474 (2021)
11. Uddin, A. H., Bapery, D., Arif, A. S. M. Depression analysis of bangla social media data using gated recurrent neural network. In: 2019 1st International Conference on Advances in Science, Engineering and Robotics Technology (ICASERT) (pp. 1–6). IEEE. (2019, May)
12. Lin, C. et al.: SenseMood: depression detection on social media. In: Proceedings of the 2020 International Conference on Multimedia Retrieval (pp. 407–411) (2020, June)
13. De Anta, L., et al.: Areas of interest and social consideration of antidepressants on English tweets: a natural language processing classification study. J. Person. Med. **12**(2), 155 (2022)
14. Kute, R.: Mental health analyzer for depression detection based on textual analysis. J. Adv. Inf. Technol. Vol, 13(1) (2022)
15. Skarpathiotaki, C.G., Psannis, K.E.: Cross-industry process standardization for text analytics. Big Data Res. **27**, 100274 (2022)

A Lightweight Siamese Network Framework for Effective Tomato Disease Recognition

Zineb Touati Hamad(✉), Mohamed Ridda Laouar, and Oumaima Saker

Laboratory of Mathematics, Informatics and Systems (LAMIS), Echahid Cheikh Larbi Tebessi University, 12002 Tebessa, Algeria
{zineb.touatihamad,ridda.laouar}@univ-tebessa.dz

Abstract. Food Security is a major global concern, and artificial intelligence (AI) is playing an increasingly crucial role in this area. In this context, automatic recognition of plant diseases, such as tomato disease detection from leaf images, is of particular importance to prevent crop losses and ensure sustainable food production. Deep learning-based tomato disease recognition methods have shown promising results, but they face major challenges. They often require large sets of labeled data, which can be expensive and time-consuming to obtain. Moreover, these traditional deep learning models consume a lot of memory and storage due to the large number of parameters they use. In this work a lightweight framework based on a Siamese network was developed for the automatic recognition of tomato leaf diseases. This lightweight framework achieved a remarkable accuracy on a subset of tomato data from the PlantVillage dataset. The experimental results demonstrate the effectiveness of this framework in handling unbalanced and small data. In addition, the backbone deep network integrated into this framework is distinguished by its lightness, which makes it far inferior to existing light deep networks.

Keywords: Food security · artificial intelligence · metric learning · Siamese neural network · tomato disease

1 Introduction

Developing countries face heightened vulnerability to the impacts of climate change, with potential repercussions such as temperature increases, altered precipitation patterns, rising sea levels, and more frequent extreme weather events. These changes pose significant threats to agriculture and food security. Among the challenges, uncontrolled plant epidemics result in substantial economic losses due to delayed disease intervention.

Establishing a predictive system for disease onset is crucial, enabling proactive measures to control outbreaks before significant crop damage occurs. This paper explores an intelligent solution leveraging Machine Learning (ML), specifically the Siamese Neural Network (SNN), to automatically detect and classify tomato plant leaves. Metric Learning is employed to discern diseases even with subtle differences in plant appearance, providing a nuanced assessment of disease levels for precise detection.

M. R. Laouar et al. (Eds.): ICISAT 2023, LNNS 981, pp. 136–148, 2024.
https://doi.org/10.1007/978-3-031-60591-8_12

The subsequent sections of this paper are structured as follows: Sect. 2 provides an overview of the Related Works. In Sect. 3, we expound upon the Materials & Methods employed in this study. Section 4 is dedicated to presenting the Results and engaging in a comprehensive discussion. Finally, Sect. 5 encapsulates the Conclusion and outlines potential perspectives for future research.

2 Related Works

The treatment of plant diseases and pests through artificial vision typically relies on conventional image processing algorithms, manually designed features, and classifiers such as SVM and K-Mean [1]. However, these traditional methods face limitations in extracting complex and detailed information from images, making disease and pest recognition challenging, often relying on experience and luck.

Traditional image recognition algorithms are built upon manually designed features, complicating their use. Additionally, these algorithms lack the capability to autonomously learn and extract features from the original image, diminishing their effectiveness in detecting plant diseases and pests.

Deep learning has emerged as an approach that overcomes the constraints of traditionally designed feature-based algorithms, gaining increasing interest among researchers. This approach has achieved significant success in various domains, including computer vision, pattern recognition, speech recognition, natural language processing, and recommendation systems.

Deep learning enables the automatic learning of features from extensive data without manual intervention. Comprising numerous layers, deep learning models possess significant capacity for self-directed learning and feature representation. They can autonomously extract features from images for tasks involving classification and recognition.

The application of deep learning in plant disease and pest recognition holds a crucial role in agriculture. Numerous deep neural network models, such as Convolutional Neural Networks (CNN), have been developed for this purpose. Tomatoes, being a major food crop susceptible to a wide range of diseases, serve as an ideal case study for deep learning in food security.

2.1 Representation Learning

The application of deep learning for plant disease and pest management involves various network architectures, such as classification, detection, and segmentation networks. This approach provides an effective and automated alternative to traditional image recognition methods for addressing plant diseases and pests (Fig. 1).

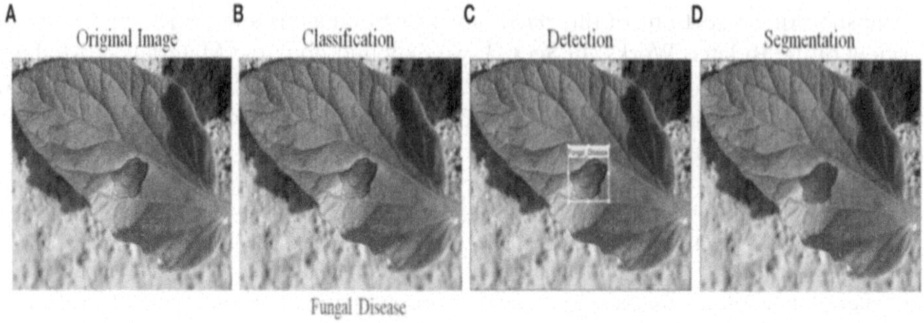

Fig. 1. Treatment of Plant Diseases: (A) raw image input, (B) classification, (C) detection, and (D) segmentation [2].

2.1.1 Classification

Recognizing plant diseases and pests in a natural environment is challenging due to significant variations in shapes, sizes, textures, colors, backgrounds, layouts, and lighting conditions of images. Convolutional Neural Networks (CNN) are widely used for classification-based on their powerful feature extraction capabilities. Established computer vision networks, including AlexNet [3], GoogleLeNet [4], VGGNet [5], ResNet [6], Inception-V4 [7], DenseNets [8], MobileNet [9] and SqueezeNet [10] are utilized for plant disease and pest classification. Classification methods can be categorized into three subcategories:

- **Using the network as a feature extractor:** Initially, researchers combined CNN's feature extraction with traditional classifiers, such as Support Vector Machines (SVM), for image classification
- **Direct use of the network for classification**: This involves directly classifying lesions using the classification network, categorized into original image classification, classification after Region of Interest (ROI) localization, and multi-category classification.
- **Using the network for lesion localization:** The classification network can also be employed for lesion localization and pixel-wise classification using techniques like sliding window, heatmap, and multi-task learning.
- **Sliding Window:** Involves sliding a small window over the original image, feeding each portion into the classification network for disease and pest detection. Classification results for each window are combined to obtain lesion localization results.
- **Heatmap:** A visual representation highlighting the relative importance of each region in an image. Darker regions indicate higher importance, and in plant disease detection, the heatmap color can indicate the probability of a region being affected by a disease or pest.
- **Multi-Task Learning Network:** Combines classification and segmentation, enhancing lesion localization precision. For instance, Ren et al. [11] built a DGVGNet model using VGNet for leaf disease identification, achieving 99.19% accuracy for

disease class identification, 94.66% for lesion segmentation, and robustness in various environments.

2.1.2 Detection

Object localization is fundamental in computer vision and closely tied to plant disease and pest detection. Deep learning-based detection methods fall into two categories: two-stage networks (e.g., Faster R-CNN [12]) and one-stage networks (e.g., SSD [13] and YOLO [14]).

- **Two-stage detection based on Faster R-CNN:** This method includes acquiring the feature map of the input image through the base network, determining anchor box confidence using the Region Proposal Network (RPN), and refining initial detection outcomes through ROI pooling.

  Fuentes et al. [15] employed Faster R-CNN for the direct localization of tomato diseases and pests. They coupled it with advanced deep feature extractors such as VGG-Net and ResNet. The achieved mean Average Precision (mAP) value reached 85.98% on a dataset comprising 5000 instances of tomato diseases and pests categorized into nine groups.
- **One-stage detection based on SSD and YOLO:** These methods simplify detection by eliminating the region proposal step. Detection heads are directly added to the base network for classification and regression, significantly speeding up inference. SSD uses VGG16 as the network trunk, while YOLO treats object detection as a regression problem, providing efficient end-to-end object detection.

Liu J, Wang X [16] used YOLO V3 to detect tomato diseases and pests, achieving a detection accuracy of 92.39% with a detection time of only 20.39 ms.

Two-stage detection models are often preferred for precise detection, while one-stage models are favored for speed. It's crucial to note that detection networks cannot replace classification networks; they complement each other in addressing plant disease and pest problems.

2.1.3 Segmentation

Segmentation transforms disease and pest detection into semantic segmentation, offering detailed information on lesions and normal areas. Two commonly used segmentation networks are Fully Convolutional Networks (FCN) [17, 18] and Mask R-CNN [19].

- **FCN:** Semantic segmentation relies on FCN, a fully convolutional neural network that extracts and encodes input image features through convolution, gradually restoring the feature image to the input image size using deconvolution or sampling.
- **Mask R-CNN:** This multitask learning method combines detection and segmentation in a single network. Wang et al. [20] used Faster R-CNN and Mask R-CNN to identify tomato diseases, achieving a high detection rate of 99.64% for all tomato disease classes.

Segmentation provides detailed information compared to classification and detection approaches but requires substantial annotated data. Additionally, pixel-wise annotations can be a costly and time-consuming process.

2.2 Metric Learning

Metric learning aims to address challenges in leaf disease recognition, such as class imbalance and limited availability of training data. It employs a lightweight architecture to achieve accurate results while utilizing fewer computational resources.

Only one study, conducted by [21], utilized a database of tomato leaf images from the PlantVillage database, encompassing a vast array of images depicting various plant diseases. The approach involved training the model using pairs of images, each pair consisting of a healthy tomato leaf image and an image of a diseased tomato leaf. The model employed a siamese network to extract features from images and measure the similarity between image pairs.

Experimental results demonstrated that the framework proposed by [21] achieved an impressive accuracy of 96.97% in tomato leaf disease recognition. Moreover, it exhibited superior performance compared to other existing approaches while being more computationally lightweight (Table 1).

Table 1. Related works

Article	Dataset	Type of Treatment	Method	Objective	Result
[11]	Private	Classification	VGGNet	Identify plant leaf diseases easily disturbed by shadows, occlusions, and light intensity	99.19%
[15]	PlantVillage	Detection	Faster R-CNN combined with VGG-Net and ResNet	Directly locate diseases and pests on tomatoes	85.98%
[16]	Private	Detection	YOLO	Used to detect diseases on tomatoes and insect pests	92.39%
[20]	Private	Detection and Segmentation	Faster R-CNN, Mask R-CNN	Identify disease class and detect/segment the location and shape of the infected area	99.64%
[21]	PlantVillage	Recognition	SNN Based CNN	The model uses a siamese neural network to extract features from images and measures the similarity between image pairs	96.97%

In the realm of plant disease detection, several challenges hinder the seamless application of deep learning methods. A pervasive issue is the constraint of small datasets, impeding the efficacy of deep learning models. Data amplification and synthesis play pivotal roles in enhancing detection efficiency, employing techniques such as mirroring, rotation, and contrast adjustment. Transfer learning, a valuable tool, aids in adapting large generic datasets to specialized domains with limited data.

The challenge of identifying small lesions in early stages is compounded by background noise, necessitating attention mechanisms for improved target detection. Fine identification of diseases and pests encounters difficulties due to varying visual features within the same class, influenced by external factors such as irregular lighting and occlusion. Moreover, detection performance is impacted by lighting variations and occlusion issues, contributing to reduced accuracy.

Addressing the challenges of real-world plant disease detection, siamese networks emerge as a potent solution. These networks, particularly in lightweight architectures, showcase impressive performance even with small and imbalanced datasets. Harnessing the benefits of siamese architectures provides a promising framework for overcoming these challenges in plant disease detection [21].

Metric learning using a siamese network proves to be a powerful tool for plant disease detection, surpassing traditional methods in detection, segmentation, and classification. Detection methods typically struggle with identifying specific regions of an image corresponding to diseased plants, especially when the disease only affects a small part of the image. Segmentation methods aim to separate diseased areas from healthy ones, but this is challenging when diseases manifest as subtle color or texture differences. Classification methods attempt to categorize the entire image as healthy or diseased based on predetermined criteria, which can be challenging with limited data or various disease types. In contrast, a siamese metric learning network can learn a similarity metric between images, enabling it to identify diseases even when the appearance of the affected plant is slightly different. By comparing images of diseased plants to a library of images of healthy plants, the network can identify specific differences indicating disease and provide a nuanced assessment of disease severity. Siam Networks (SNN) can be trained with limited data, a common challenge in plant disease detection. Utilizing transfer learning, the network can be trained on a large dataset of plant images and then fine-tuned on a smaller dataset of disease images, overcoming the challenge of limited data and improving disease detection accuracy.

In conclusion, while detection, segmentation, and classification methods have their utility in plant disease detection, a siamese network for metric learning offers several advantages in terms of accuracy, versatility, and ease of training. By learning a similarity metric between images, siamese networks can effectively identify diseases even when the appearance of the affected plant is only slightly different, making them an efficient tool for plant disease detection.

Recent lightweight siamese architectures have demonstrated exceptional performance with small and imbalanced datasets. Therefore, developing a comprehensive framework by harnessing the advantages of the siamese architecture is crucial as the first step to overcome these challenges.

3 Materials and Methods

The architecture of our system comprises dataset handling, data preprocessing to prepare images, modeling to construct a disease detection model, and evaluation to measure the model's performance on a separate test set (Fig. 2).

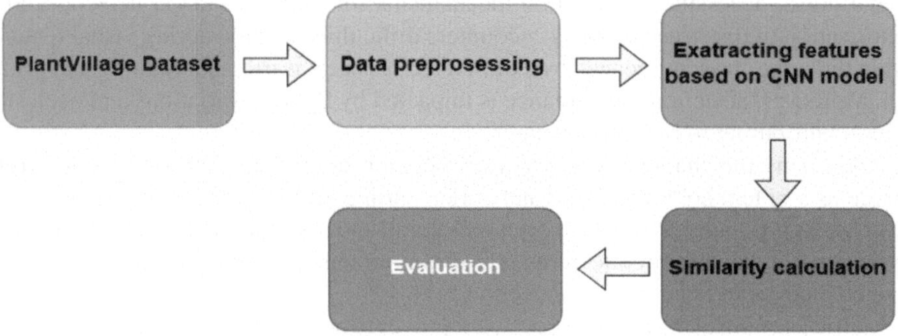

Fig. 2. Proposed system architecture

3.1 Dataset Description

We extracted leaf images from PlantVillage [22], a recent and open database. This database comprises a total of 54,306 images from 14 different crops, covering 38 disease classes across various plant types. The dataset specifically includes 14,529 images sampled from nine (09) disease classes known to affect tomato plants, in addition to a class for healthy leaves. The original dataset images are sized at 224 × 224 pixels. In our study, we focused solely on the tomato crop within this broader dataset. Table 2 presents the description of the Tomato plants dataset, while Fig. 3 showcases the different diseases of tomato plants.

Table 2. PnatVillage Tomato image dataset description

Dataset	Total images	Plant species	Classes	Tomato leaf images	Tomato classes
Plantvillage	54,306	14	38	16,578	10

Fig. 3. Examples of different diseases of tomato plants. A) Healthy leaf B) Early Blight C) Late Blight D) Septoria Leaf Spot E) Yellow Leaf Curl Virus F) Bacterial Spot G) Target Spot H) Spider Mite I) Leaf Mold J) Mosaic Virus

3.2 Data Preprocessing

All images used in the database were already resized to a fixed size of 112 × 112 pixels, with 50 samples utilized, each in grayscale for every class. Subsequently, the data was divided into two parts: 75% for training data and 25% for test data from the dataset.

3.3 Proposed Method

Our proposed model, a Siamese Neural Network (SNN), is a specialized neural network designed for measuring the similarity or distance between two input images. Specifically, these inputs consist of a pair of images: the first referred to as the "anchor" and the second as the "positive." The primary objective is to quantify the Euclidean distance between corresponding points within the feature space, facilitated by the use of a Convolutional Neural Network (CNN).

Following a Siamese architecture, our SNN model employs shared weights and layers from the CNN, ensuring consistent feature extraction from both images. This implies that both the anchor and positive images undergo the same CNN processing to extract their respective features.

After the Convolutional Neural Network (CNN) finishes extracting features from both images, the computation of the Euclidean distance between these features takes place. The Euclidean distance acts as a metric to measure the similarity between two points in a multidimensional space. It is calculated by determining the square root of the sum of the squares of differences in the coordinates of the points.

Leveraging the Euclidean distance enables us to evaluate the degree of similarity or dissimilarity between the two images. A smaller Euclidean distance indicates a higher level of similarity, while a larger distance signifies greater dissimilarity. Therefore, the SNN model, integrated with a CNN, facilitates the comparison of extracted features from images and quantifies their similarity by assessing the euclidean distance between corresponding points within the feature space (Fig. 4).

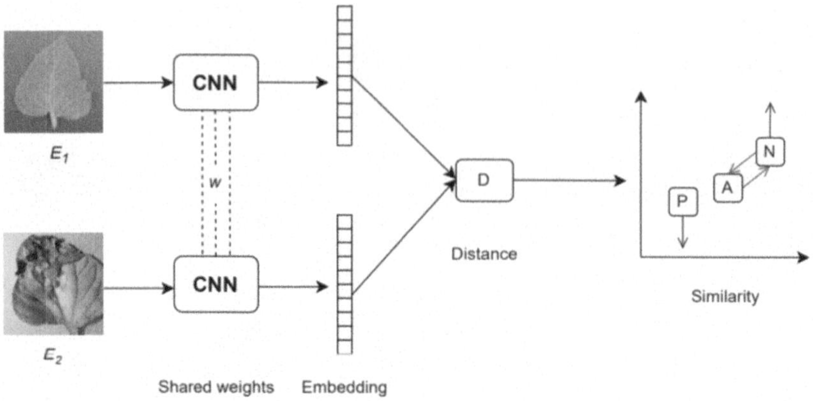

Fig. 4. Structure of the tomato disease recognition framework. The input images $\boldsymbol{E}_1$ and $\boldsymbol{E}_2$ are processed through the joint weights (W) in the CNN for the Siamese network. After that, the estimated distance (D) is calculated as the Euclidean distance between the CNN ($\boldsymbol{E}_1$) and CNN ($\boldsymbol{E}_2$) outputs. The similarity then is subsequently determined based on D.

Deep Convolutional Neural Networks (CNNs) are widely acknowledged and recommended for addressing classification problems. In alignment with this, Fig. 5 details our proposed CNN architecture. Within the architecture, the convolutional layers play a pivotal role. The first conv2D layer is specifically designed for identity extraction, employing a convolution operation with 16 filters of size 3 × 3. This process is applied to filter an input image with dimensions of 112*112*1, utilizing the Rectified Linear Unit (ReLU) activation function. The second convolutional layer, positioned third in the model, shares substantial similarities with the initial layer. It takes as input the output of the first convolutional layer after a pooling operation with 32 filters of size 3 × 3. The third convolutional layer, occupying the fifth position in the model, employs 16 filters of size 3 × 3 with the application of the ReLU activation function. Furthermore, the architecture incorporates two max-pooling layers to effectively reduce the spatial dimension (feature map) of the convolutional layers. All three pooling layers utilize the same filter size of 2 × 2. The model concludes with a Flatten layer and three Fully-Connected layers, each integrating the ReLU activation function, followed by a dropout to mitigate the challenges associated with overfitting. This comprehensive CNN architecture is tailored to optimize performance in classification tasks.

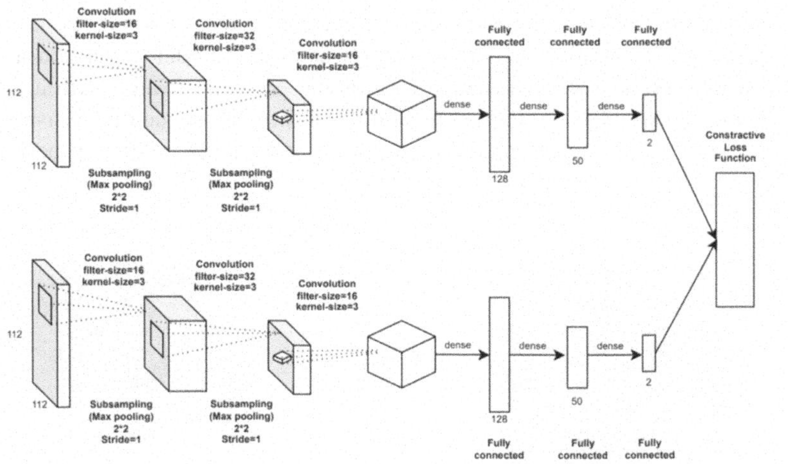

Fig. 5. The schematic representation of the CNN serving as the core of the proposed framework.

4 Results and Discussion

This section presents the results obtained during the training phase of the proposed model after its compilation. The first graph depicts the accuracies for 75% training data and 25% validation data across the number of epochs. It can be observed that the accuracy reaches a high level of 99.73% after 20 epochs.

The second graph illustrates the error percentages (Loss) for 75% training data and 25% validation data. It is noticeable that these percentages decrease rapidly as the model's accuracy increases, which is entirely normal during the training phase. As the accuracy improves, the classification error diminishes rapidly (Fig. 6).

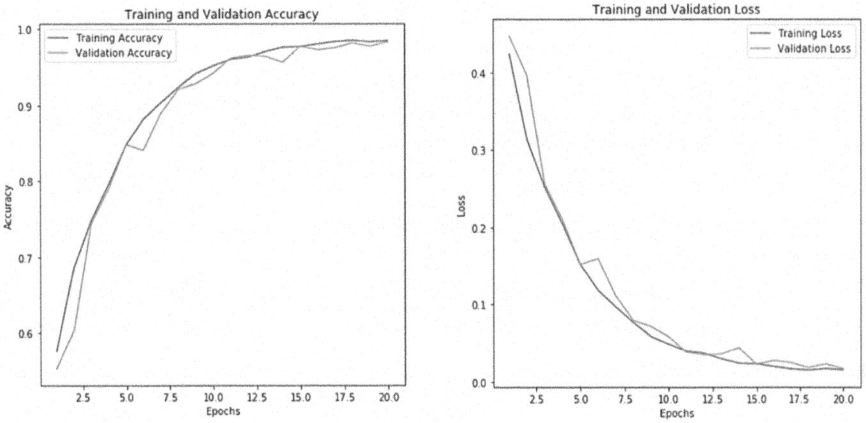

Fig. 6. Model Accuracy and loss

Once our model is validated and ready for execution, we can input two images and receive, as a system output, a prediction report using the model trained on a specific pair of test images. The system then compares the prediction value with a threshold (0.5) to determine whether the images are similar or not. Subsequently, it displays both the ground truth value and the model prediction to assess the prediction performance (Fig. 7).

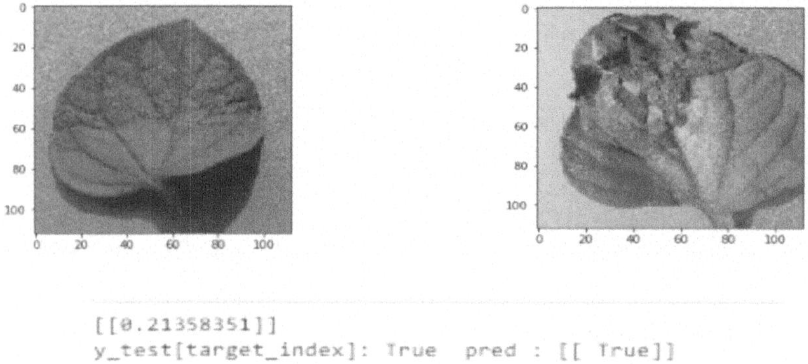

Fig. 7. Model exit states

The result [0.21358351] corresponds to the prediction value returned by the model. This value represents the probability that the two test images are similar, where a value closer to 1 indicates a higher likelihood of similarity.

Next, y_test[target_index]: True indicates the ground truth value for this specific example, meaning that the ground truth indicates that the two test images are similar.

Finally, pred: [[True]] indicates the model's prediction after applying a threshold of 0.5. In this case, the prediction value is higher than 0.5, corresponding to a prediction of similarity between the two images. Therefore, the model correctly predicts the similarity between the images, as indicated by the ground truth.

In summary, the model successfully predicts the similarity between the two test images because the prediction value is above the threshold (0.5) and corresponds to the ground truth (True) indicating that the images are similar (Table 3).

Table 3. Evaluation of current approaches for recognizing diseases in tomato leaves.

	Dataset	Architecture	Backbon model	Epochs	Accuracy
[21]	PlantVillage	SNN	CNN	10	96.97%
Proposed Study	PlantVillage	SNN	CNN	10	**99.73%**

5 Conclusion and Perspectives

In conclusion, the SNN architecture, enriched by a CNN model for leaf image similarity, presents a promising breakthrough in the domain of plant disease detection. Its remarkable precision in identifying diseases, even amidst subtle variations in the appearance of afflicted tomato plants, underscores its efficacy. The incorporation of a CNN model empowers the SNN architecture to adeptly capture distinctive features of tomato leaves, facilitating robust image comparison and similarity assessment. This resilience to minor variations in leaf appearance is pivotal for the early detection of diseases, especially those manifesting subtle symptoms in their initial stages. The model's proficiency in identifying tomato plant diseases, even in the face of slight appearance differences, positions it as a valuable tool for both farmers and researchers. Early and accurate disease detection not only enables swift preventive measures but also minimizes crop losses, optimizing overall crop management strategies. Crucially, this architecture holds the potential for extension to other plant species, offering a versatile solution for disease detection across a wide range of crops. This adaptability contributes to the efficiency and sustainability of agricultural practices.

Looking forward, several promising avenues emerge. First and foremost, improving the quality and diversity of training data stands out as a crucial step in refining the model. This enhancement can further augment the model's capacity to recognize various disease patterns, thereby boosting overall performance. Moreover, transitioning the model from laboratory settings to real-world agricultural applications is paramount. Collaborating with agricultural experts can provide valuable insights, ensuring the practicality and effectiveness of the system in the field. This ongoing collaboration and feedback loop will enhance the model's adaptability to real-world challenges and specific agricultural contexts, ensuring its successful application in diverse farming scenarios.

Acknowledgment. This research has received support from the Algerian General Direction of Scientific Research and Technological Development (DGRSDT) and the LAMIS Laboratory.

References

1. Harakannanavar, S.S., Rudagi, J.M., Puranikmath, V.I., Siddiqua, A., Pramodhini, R.: Plant leaf disease detection using computer vision and machine learning algorithms. Glob. Transit. Proc. **3**, 305–310 (2022). https://doi.org/10.1016/j.gltp.2022.03.016
2. Shoaib, M., et al.: An advanced deep learning models-based plant disease detection: a review of recent research. Front. Plant Sci. **14**, 1158933 (2023)
3. Alom, M.Z. et al.: The history began from AlexNet: a comprehensive survey on deep learning approaches. http://arxiv.org/abs/1803.01164 (2018). https://doi.org/10.48550/arXiv.1803.01164
4. Yuesheng, F., et al.: Circular fruit and vegetable classification based on optimized GoogLeNet. IEEE Access. **9**, 113599–113611 (2021). https://doi.org/10.1109/ACCESS.2021.3105112
5. Muhammad, U., Wang, W., Chattha, S.P., Ali, S.: Pre-trained VGGNet architecture for remote-sensing image scene classification. In: 2018 24th International Conference on Pattern Recognition (ICPR), pp. 1622–1627 (2018). https://doi.org/10.1109/ICPR.2018.8545591

6. Targ, S., Almeida, D., Lyman, K.: Resnet in resnet: generalizing residual architectures, http://arxiv.org/abs/1603.08029 (2016). https://doi.org/10.48550/arXiv.1603.08029
7. Szegedy, C., Ioffe, S., Vanhoucke, V., Alemi, A.: Inception-v4, Inception-ResNet and the impact of residual connections on learning. In: Proc. AAAI Conf. Artif. Intell. vol. 31 (2017). https://doi.org/10.1609/aaai.v31i1.11231
8. Zhang, K., Guo, Y., Wang, X., Yuan, J., Ding, Q.: Multiple feature reweight densenet for image classification. IEEE Access. **7**, 9872–9880 (2019). https://doi.org/10.1109/ACCESS.2018.2890127
9. Zaki, S.Z.M., Zulkifley, M.A., Stofa, M.M., Kamari, N.A.M., Mohamed, N.A.: Classification of tomato leaf diseases using MobileNet v2. IAES Int. J. Artif. Intell. **9**, 290 (2020)
10. Hidayatuloh, A., Nursalman, M., Nugraha, E.: Identification of tomato plant diseases by leaf image using squeezenet model. In: 2018 International Conference on Information Technology Systems and Innovation (ICITSI). pp. 199–204 (2018). https://doi.org/10.1109/ICITSI.2018.8696087
11. Shougang, R., Fuwei, J., Xingjian, G., Peishen, Y., Wei, X., Huanliang, X.: Deconvolution-guided tomato leaf disease identification and lesion segmentation model. J. Agric. Eng. **36**, 186–195 (2020)
12. Ren, S., He, K., Girshick, R., Sun, J.: Faster R-CNN: towards real-time object detection with region proposal networks. In: Advances in Neural Information Processing Systems. Curran Associates, Inc. (2015)
13. Liu, W., Anguelov, D., Erhan, D., Szegedy, C., Reed, S., Fu, C.-Y., Berg, A.C.: SSD: Single Shot MultiBox Detector. In: Leibe, B., Matas, J., Sebe, N., Welling, M. (eds.) Computer Vision – ECCV 2016. Springer International Publishing, Cham (2016). https://doi.org/10.1007/978-3-319-46448-0_2
14. Jiang, P., Ergu, D., Liu, F., Cai, Y., Ma, B.: A review of Yolo algorithm developments. Proc. Comput. Sci. **199**, 1066–1073 (2022). https://doi.org/10.1016/j.procs.2022.01.135
15. Fuentes, A., Yoon, S., Kim, S.C., Park, D.S.: A robust deep-learning-based detector for real-time tomato plant diseases and pests recognition. Sensors. **17**, 2022 (2017). https://doi.org/10.3390/s17092022
16. Liu, J., Wang, X.: Tomato diseases and pests detection based on improved Yolo V3 convolutional neural network. Front. Plant Sci. **11**, 898 (2020)
17. Khan, S., Narvekar, M.: Disorder Detection in Tomato Plant Using Deep Learning. In: Vasudevan, H., Michalas, A., Shekokar, N., Narvekar, M. (eds.) Advanced Computing Technologies and Applications. Springer, Singapore (2020). https://doi.org/10.1007/978-981-15-3242-9_19
18. Long, J., Shelhamer, E., Darrell, T.: Fully convolutional networks for semantic segmentation. In: Presented at the Proceedings of the IEEE Conference on Computer Vision and Pattern Recognition (2015)
19. He, K., Gkioxari, G., Dollar, P., Girshick, R.: Mask R-CNN. In: Presented at the Proceedings of the IEEE International Conference on Computer Vision (2017)
20. Wang, Q., Qi, F., Sun, M., Qu, J., Xue, J.: Identification of tomato disease types and detection of infected areas based on deep convolutional neural networks and object detection techniques. Comput. Intell. Neurosci. **2019**, e9142753 (2019). https://doi.org/10.1155/2019/9142753
21. Thuseethan, S., Vigneshwaran, P., Charles, J., Wimalasooriya, C.: Siamese network-based lightweight framework for tomato leaf disease recognition. http://arxiv.org/abs/2209.11214 (2022). https://doi.org/10.48550/arXiv.2209.11214
22. Hughes, D.P., Salathe, M.: An open access repository of images on plant health to enable the development of mobile disease diagnostics. http://arxiv.org/abs/1511.08060 (2016). https://doi.org/10.48550/arXiv.1511.08060

Machine Learning Algorithms for Hospital Length of Stay Prediction

Zineb Touati Hamad(✉), Mohamed Ridda Laouar, and Gadri Dhouha

Laboratory of Mathematics, Informatics and Systems (LAMIS), Echahid Cheikh Larbi Tebessi University, 12002 Tebessa, Algeria

{zineb.touatihamad,ridda.laouar}@univ-tebessa.dz

Abstract. Efficient prediction of hospital length of stay is crucial for optimizing healthcare operations, resource allocation, and cost-effectiveness. This research harnesses artificial intelligence, specifically machine learning techniques, to enhance the accuracy of predicting hospital Length of Stay. The principal objective is to assist healthcare professionals in selecting the most suitable prediction method based on their specific requirements and constraints. Customized machine learning models have demonstrated success in achieving precise predictions, streamlining the acquisition of patient data, and evaluating both hospitalization and treatment durations. After evaluating various algorithms, the Random Forest algorithm emerged as the most effective for predicting hospital stay durations, yielding a Root Mean Square Error (RMSE) of 0.24 and a coefficient of determination (R^2) of 0.94. These compelling findings underscore the efficacy of the Random Forest algorithm in accurately forecasting the duration of hospital stays.

Keywords: Hospital · length of stay · prediction · machine learning

1 Introduction

The digitization of medical information stands as a pivotal process in the modernization of healthcare, transforming patient records into electronic files for enhanced storage, sharing, and utilization across diverse healthcare facilities. This shift aims to improve healthcare quality while mitigating costs associated with managing traditional paper medical records. Leveraging large language models in this context has become a topic of interest, particularly in the realm of predicting hospital Length of Stay (LoS) through the application of artificial intelligence (AI).

In recent years, AI has made significant strides in various domains, including healthcare. One specific application of AI in healthcare is the prediction of Hospital LoS a critical aspect for healthcare professionals, patients, and healthcare institutions. AI, as an advanced technology, allows machines to learn, reason, and make autonomous decisions, offering a unique potential for predicting hospital stay durations more accurately than traditional methods.

M. R. Laouar et al. (Eds.): ICISAT 2023, LNNS 981, pp. 149–163, 2024.
https://doi.org/10.1007/978-3-031-60591-8_13

This paper explores the intersection of AI and the prediction of hospital LoS, acknowledging the limitations of traditional statistical models and embracing the transformative capabilities of AI. The integration of AI in predicting hospital stay durations brings value by analyzing vast amounts of real-time data, detecting complex patterns, and ultimately enhancing the precision of predictions.

The subsequent sections of this paper are structured in the following manner: Sect. 2 delineates the problem, providing a comprehensive description. In Sect. 3, we present an overview of the current state of the art in the field. Section 4 provides a detailed exposition of the Materials & Methods employed in this study. Subsequently, Sect. 5 encapsulates the Discussion and Comparison of the obtained results. The paper culminates in Sect. 6, wherein the findings and insights are synthesized, leading to the conclusive remarks of this work.

2 Problem Description

Optimizing hospital resources emerges as a critical imperative in meeting the ever-growing healthcare demands of our population. Amidst this challenge, the duration of hospital stays stands out as a pivotal metric, reflecting the efficiency of healthcare systems. However, this scenario presents a thrilling opportunity for innovation. Our study delves into a compelling problem: can artificial intelligence (AI) revolutionize the precision of predicting hospital LoS? With healthcare professionals grappling with escalating workloads, the imperative to understand how AI functions in this context becomes even more exciting. The research not only seeks to address this urgent question but also aims to inject vitality by exploring the superiority of AI over traditional prediction methods. It ventures into the dynamic realm of identifying the most effective AI method and meticulously selecting datasets for accurate predictions. As we embark on this scientific journey, our goal is not only to provide insightful answers but to paint a vivid picture of the transformative conditions under which AI becomes a beacon, significantly reshaping resource management and uplifting the quality of patient care.

3 State of the Art

The realm of predicting hospital LoS has traditionally relied on classical statistical models, often hindered by the intricacies of health data and the variability of risk factors. However, in our quest for precision, artificial intelligence (AI) emerges as a captivating force, capable of real-time analysis of vast datasets and deciphering intricate patterns to elevate prediction accuracy.

3.1 Classical Methods for LoS Prediction

Within the classical landscape, predictive methodologies hinge on statistical models utilizing available clinical data such as age, gender, diagnosis, medical history, prior treatments, and laboratory/imaging results. While these approaches may differ across healthcare institutions and countries, a spectrum of widely adopted methods prevails [1].

A. Rule-Based Methods
Anchored in pre-defined clinical rules, these methods predict hospital stay duration based on the patient's diagnosis, treatments, and medical history. Crafted by physicians or healthcare professionals, these rules draw upon clinical experience and expertise.
B. Expertise-Driven Approaches
Here, clinical decision-making is entrusted to healthcare professionals who leverage their experience and expertise. Physicians and healthcare practitioners exercise clinical judgment to predict hospital stay duration based on the patient's health status and their cumulative experience.

It's essential to acknowledge the inherent limitations of these classical methods, often lacking the precision required for robust hospital LoS prediction. Consequently, the burgeoning field of AI in predicting LoS is gaining momentum, emerging as a promising alternative to overcome the shortcomings of traditional approaches. As we embark on this exploration, the fusion of AI and healthcare promises not just innovation, but a paradigm shift in the accuracy and reliability of predicting LoS.

3.2 Intelligent Methods for LoS Prediction

Venturing into the realm of intelligent methodologies for predicting length of hospital stay, we harness the prowess of Artificial Intelligence (AI) to decipher patient outcomes based on available clinical data. The excitement lies in exploring cutting-edge techniques that not only enhance prediction accuracy but also revolutionize decision-making in hospital stay management.

A. Artificial Neural Networks (ANNs)
Artificial Neural Networks stand out as a formidable intelligent method widely employed in DDSH prediction. Inspired by the human brain, these networks utilize machine learning models to uncover intricate relationships between patient characteristics and LoS. While requiring substantial training data and modeling expertise, ANNs excel in capturing non-linear relationships, empowering precise predictions and advancing decision-making in hospital stay management [2].
B. Decision Trees
Intelligence takes root in Decision Trees, presenting a straightforward and interpretable method for hospital LoS prediction. Leveraging patient characteristics, decision trees make intuitive and interpretable predictions, offering valuable insights into factors influencing hospital stay duration. Whether used independently or in conjunction with other techniques, decision trees provide an intuitive approach, enriching the understanding of hospital LoS influencers [3].
C. Random Forests
Enter the enchanting world of Random Forests, a prevalent intelligent method for DDSH prediction. By amalgamating predictions from multiple decision trees, Random Forests craft a more accurate final prediction. Proficient in handling vast datasets, providing variable importance measures, and balancing precision with interpretability, they emerge as a potent tool for hospital LoS prediction, promising elevated accuracy and insightful interpretability [4].

D. Support Vector Machines (SVMs)
Support Vector Machines (SVMs) stake their claim as another intelligent method in hospital LoS prediction. As a branch of effective supervised learning algorithms, SVMs, particularly Support Vector Regression (SVR), excel in minimizing prediction errors by seeking an optimal function fit to training data. Utilizing support vectors to determine this function, SVMs construct an optimal decision boundary, showcasing their prowess in predicting hospital stay duration [5].

3.3 Related Works on Hospital LoS Prediction

Exploring the landscape of predictive analytics for LoS, a plethora of prior studies have illuminated the path, offering insights into leveraging Machine Learning techniques for enhanced prognostication. These endeavors, akin to stepping stones, collectively contribute to the evolving tapestry of knowledge in the realm of healthcare management.

A. Machine Learning Models for Hospital Stay Prediction
In a comparative study, the application of Random Forest (RF) and Gradient Boosting (GB) algorithms on Microsoft's "Lengthofstay.csv" dataset showcased the superiority of RF in predicting hospital stay duration, as evidenced by superior Mean Absolute Error (MAE) performance [6].
B. Predicting ICU Patient Duration and Mortality
Addressing the imperative of reducing patient mortality in Intensive Care Units (ICUs), researchers employ Machine Learning methods such as Decision Trees (DT), Random Forests (RF), Artificial Neural Networks (ANN), Bayesian networks, and Support Vector Machines (SVM). Their comprehensive review, based on MIMIC II dataset, delves into predicting measurable outcomes like complication risk, mortality, and hospital stay duration [4].
C. Data Exploration Techniques for Predicting Cardiac Patient Stay
Utilizing data exploration techniques, this study forecasts the DDSH for patients with coronary diseases. Employing decision trees, Support Vector Machines (SVM), and Artificial Neural Networks (ANN) on a dataset of 4,948 patients, the SVM algorithm emerges as the most accurate, achieving a precision of 96.4% [2].
D. Neural Networks for Hospital LoS Prediction
Harnessing the neuralnet package in R, researchers train a model based on demographic features from MIMIC III database. The model demonstrates an accuracy of approximately 79% in predicting whether a patient's stay in the Intensive Care Unit (ICU) would be short (≤ 5 days) or long (> 5 days) after departure [7].
E. ML-Based Patient Stay Prediction at Admission
Examining the feasibility of predicting patient stay duration at admission, this work utilizes various Machine Learning algorithms. The results suggest the adaptability of these algorithms in broader hospital settings, emphasizing the potential for generalized application [8].
F. Prediction of Prolonged Hospital Stay after Surgery
Developing a predictive model for prolonged hospital stays post-surgery, this study employs supervised learning techniques. The Random Forest method emerges as the most accurate and stable prediction model, underscoring the utility of supervised learning in patient prognosis and safety enhancement [9].

G. Neural Network Model for Cholecystitis Patients

Applying a Backpropagation Neural Network (BP) model, this study aims to predict hospital stay duration for patients with cholecystitis. The model achieves an accuracy of around 80%, emphasizing the significance of stay duration assessment in evaluating medical costs and resource utilization [10].

The state-of-the-art approaches in LoS predominantly involve intelligent methodologies such as Artificial Neural Networks (ANNs) and ensemble learning techniques like Random Forests. While these methods have shown considerable success, they are not without challenges and limitations. One significant challenge is the interpretability of complex models like ANNs, making it difficult for healthcare professionals to comprehend and trust the predictions fully. Additionally, ANNs often require substantial amounts of training data and specialized expertise.

In contrast, Decision Trees provide a transparent and interpretable method, offering insights into factors influencing LoS. However, Decision Trees may suffer from overfitting or oversimplification, potentially compromising their predictive accuracy in intricate healthcare scenarios.

Support Vector Regression (SVR), another intelligent method, is effective in minimizing prediction errors. However, SVR's performance can be sensitive to the choice of hyperparameters, and the interpretation of the underlying decision boundaries may not be as straightforward as with Decision Trees.

Multilayer Perceptron Regression (MLPR), inspired by the human brain, excels in capturing non-linear relationships but may face challenges related to model complexity and potential overfitting, especially in scenarios with limited data.

Considering these challenges, the rationale for incorporating Random Forests lies in their ability to address several shortcomings of individual methods. Random Forests leverage the ensemble of Decision Trees, providing a balance between interpretability and predictive power. By aggregating predictions from multiple trees, Random Forests mitigate overfitting concerns present in individual Decision Trees. Moreover, the method's capability to handle vast datasets aligns well with the complexity and variability of health data, enhancing its potential for robust LoS predictions. The combination of interpretability, ensemble learning, and adaptability to complex data makes Random Forests a compelling choice within our diverse set of predictive models, ensuring a comprehensive and accurate approach to hospital LoS prediction.

4 Materials and Methods

In the context of our theme, the measurement of hospital LoS will be calculated in terms of the number of days. This approach, measuring in days, offers a more practical perspective that aligns with our clinical reality, as hospital LoS is typically assessed and managed based on full days of hospitalization.

By using this day-based measurement scale, it becomes easier for us to plan our hospital resources, organize medical appointments and procedures, and coordinate post-hospital care.

The prediction of hospital LoS is calculated through the following steps (Fig. 1). The initial step involves collecting the necessary data from the "lengthofstay.csv"

DataSet, extracting information about hospital stays. Subsequently, data analysis, cleaning, variable selection, data preprocessing, and modeling steps are undertaken. These processes contribute to constructing LoS prediction models based on patient characteristics. The developed models are then evaluated for accuracy and performance to enhance decision-making in hospital stay management.

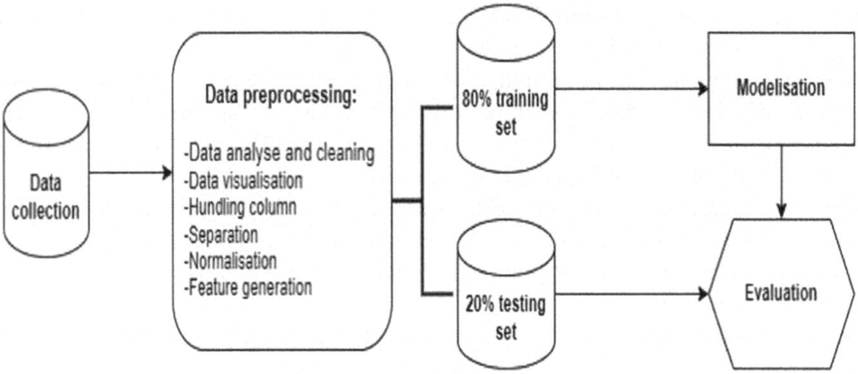

Fig. 1. Proposed framework for our prediction model.

4.1 Data Collection

The required data is collected from the "lengthofstay.csv" DataSet, containing information about hospital stays. This data has been generously provided as an open-source resource by Microsoft. It includes information on a cohort of 100,000 hospitalized patients, encompassing crucial indicators related to their health conditions and the duration of their hospital stays. The comprehensive nature of this dataset is invaluable for predicting the length of patients' hospital stays.

Notably, the "LengthOfStay.csv" file contains detailed information about each patient within this dataset [11].

This dataset serves as a foundational component for training and evaluating machine learning models designed to forecast hospital stay durations. The Table 1 provides an overview of the dataset description.

Table 1. Dataset description [11]

Index	Data Field	Type	Description
01	eid	Integer	Unique identifier for hospital admission
02	vdate	String	Visit date
03	rcount	Integer	Number of readmissions in the last 180 days
04	gender	String	Patient's gender - M or F
05	dialysisrenalendstage	String	Indicator of renal disease during the episode of care
06	asthma	String	Indicator of asthma during the episode of care
07	irondef	String	Indicator of iron deficiency during the episode of care
08	pneum	String	Indicator of pneumonia during the episode of care
09	substancedependence	String	Indicator of substance dependence during the episode of care
10	psychologicaldisordermajor	String	Indicator of major psychological disorder during the episode of care
11	depress	String	Indicator of depression during the episode of care
12	psychother	String	Indicator of other psychological disorders during the episode of care
13	fibrosisandother	String	Indicator of fibrosis during the episode of care
14	malnutrition	String	Indicator of malnutrition during the episode of care
15	hemo	String	Indicator of blood disorder during the episode of care
16	hematocrit	Float	Average value of hematocrit during the episode of care (g/dL)
17	neutrophils	Float	Average value of neutrophils during the episode of care (çells/μL)
18	sodium	Float	Average value of sodium during the episode of care (mmol/L)
19	glucose	Float	Average value of glucose during the episode of care (mmol/L)

(*continued*)

Table 1. (*continued*)

Index	Data Field	Type	Description
20	bloodureanitro	Float	Average value of blood urea nitrogen during the episode of care (mg/dL)
21	creatinine	Float	Average value of creatinine during the episode of care (mg/dL)
22	bmi	Float	Average body mass index during the episode of care (kg/m^2)
23	pulse	Float	Average pulse during the episode of care (beats/min)
24	respiration	Float	Average respiration rate during the episode of care (respirations/min)
25	secondarydiagnosisnonicd9	Integer	Indicator of whether a non-ICD-9 formatted diagnosis has been coded as a secondary diagnosis
26	discharged	String	Discharge date
27	facid	Integer	Facility identifier where the episode of care occurred
28	lengthofstay	Integer	Duration of stay for the episode of care

4.2 Data Preprocessing

- **Data Analysis and Cleaning**
  The data is analyzed to understand its structure, types of variables present, and to identify potential errors or missing values. Data is cleaned by addressing missing values, removing erroneous or incomplete records, and correcting any errors.
- **Data Visualization**
  The Fig. 2 depicts the count of individuals by gender (F for female and M for male) based on their length of stay. The length of stay is represented on the x-axis, ranging from 0 to 17 days, while the count is on the y-axis, reaching up to 12,000 persons. It is clear that the residency of females is always higher than that of males, as well as the maximum period of residency is approximately 10 days and is almost non-existent from 11 to 17 days.

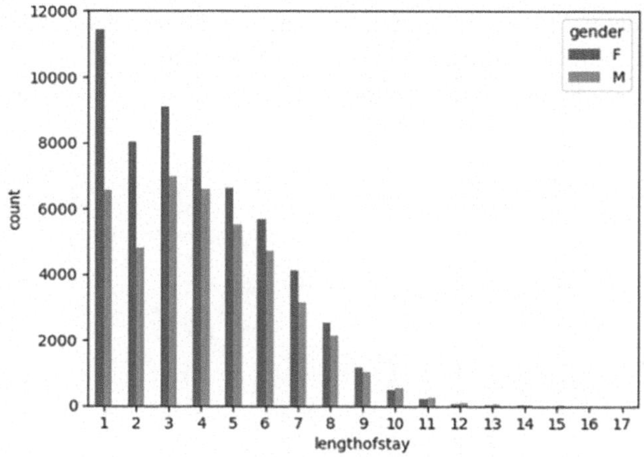

Fig. 2. Length of Stay and Gender

The Fig. 3 resresents the count of patients against their length of stay in a facility.

The x-axis, labeled as "lengthofstay," ranges from 0 to 17, and the y-axis, labeled as "count," ranges from 0 to 17500. Different colors represent the readmission count. A significant number of patients (over 15000) had a length of stay of one day and were not readmitted.

There are smaller bars representing other lengths of stays and readmissions, with the second highest being patients who stayed for two days and were not readmitted.

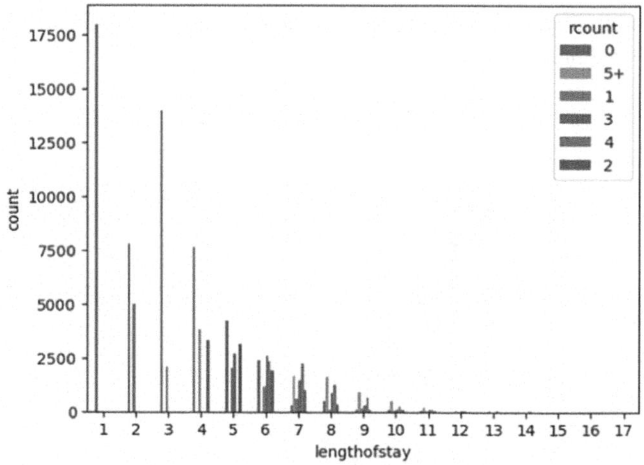

Fig. 3. Length of Stay and Readmission

The Fig. 4 depicts the count of patients with and without iron deficiency during their episode of care, categorized by the length of stay in days. The blue bars represent patients without iron deficiency (irondef = 0), and the orange bars represent patients with iron deficiency (irondef = 1). As the lengthofstay increases, both categories decrease in count but always more patients without iron deficiency than with it.

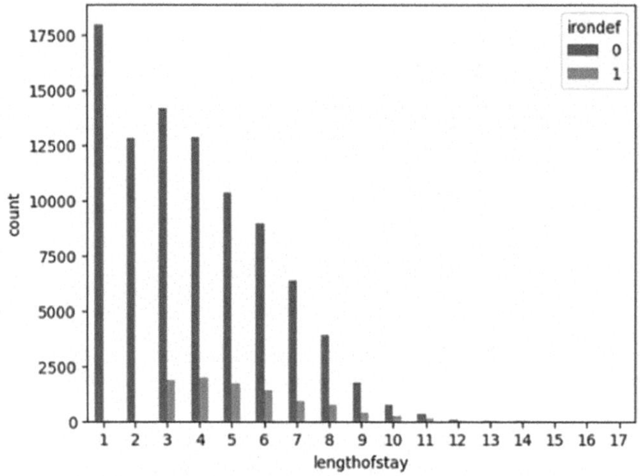

Fig. 4. Length of Stay and Iron Deficiency

The Fig. 5 is a bar graph representing counts of individuals with and without depression based on their length of stay.

The x-axis is labeled "lengthofstay" and ranges from 0 to 17, and the y-axis is labeled "count" and ranges from 0 to 17500. There are two sets of bars, one in blue representing individuals without depression (depress = 0) and one in orange representing individuals with depression (depress = 1).

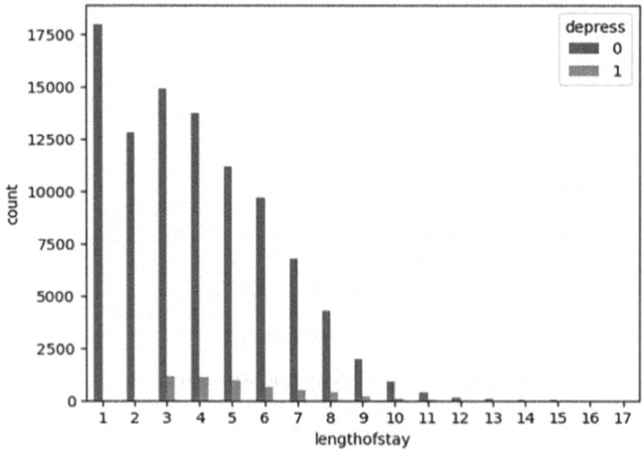

Fig. 5. Length of Stay and Depression

The majority of the counts are for individuals without depression, there are significantly fewer individuals with depression appear in only the middle length of stay (Between 3 to 10 days).

- **Handling Columns**
  Managing columns involves various operations, including handling categorical columns, encoding specified columns in the dataset, and creating new binary columns to represent each unique value.
- **Separation of Target and Feature Variables**
  Variables in the dataset are separated into target variables (dependent) and feature variables (independent) to facilitate the modeling process.
- **Normalization of Numerical Variables**
  Numeric variables are normalized using the "StandardScaler" class, ensuring all variables are brought to a common scale. This normalization aids in the learning process of predictive models.
- **Feature Removal**
  To enhance the predictive model for hospital stay duration, certain columns such as discharge date and visit date, which directly reflect the length of stay, are excluded. By excluding these features, the model is encouraged to discover additional variables influencing the duration of stay beyond explicit date information.
- **Creation of a New Feature**
  A new feature named "total number of issues" is created by summing binary columns such as dialysisrenalendstage, asthma, irondef, pneum, substancedependence, psychologicaldisordermajor, depress, psychother, fibrosisandother, malnutrition, and hemo. This feature condenses information into a single measure, capturing the overall count of issues for each data point. This simplifies the model, provides insights into the extent of present issues, and reduces data dimensionality.
- **Data splitting**
  The models are trained on a subset of the data, utilizing 80% for training (train) and 20% for testing. A random state of 42 ensures consistent division into training and testing sets each time the code is executed.

4.3 Modelisation

Various prediction algorithms, including Linear Regression (MLPR), Support Vector Machines for Regression (SVM-SVR), Decision Trees, or Random Forests, are employed to construct hospital stay duration prediction models.

The models are trained using the training dataset to learn patterns and relationships within the data. Predictions are generated using the trained models on the test dataset, providing insights into how well the models generalize to new, unseen data. The performances of each model are visualized to compare their effectiveness in predicting hospital stay durations.

4.4 Evaluation

Models undergo evaluation using performance metrics such as mean error, coefficient of determination (R^2), or Root Mean Squared Error (RMSE). These metrics aid in selecting the most effective model for predicting hospital LoS. Table 2 displays the performance results obtained for each model.

Table 2. Models performance.

Model	RMSE	R^2
Random Forest	**0.24***	**0.94***
Decision Tree	0.34	0.87
SVR	0.30	0.90
MLPR	0.26	0.93

Figure 6 displays the results of each model along with its RMSE. A lower RMSE indicates better performance, as it signifies that the model's predictions are closer to the actual values.

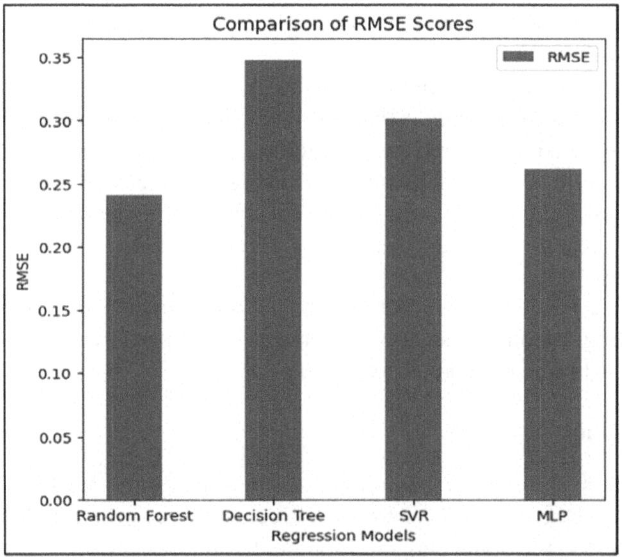

Fig. 6. Comparison of RMSE scores.

Figure 7 displays the results of each model along with its R^2. A high R^2 indicates that the model can explain a large portion of the variance in the target variable, suggesting a good fit of the model to the data. This can be interpreted as a robust performance of the model in capturing variations in the target variable.

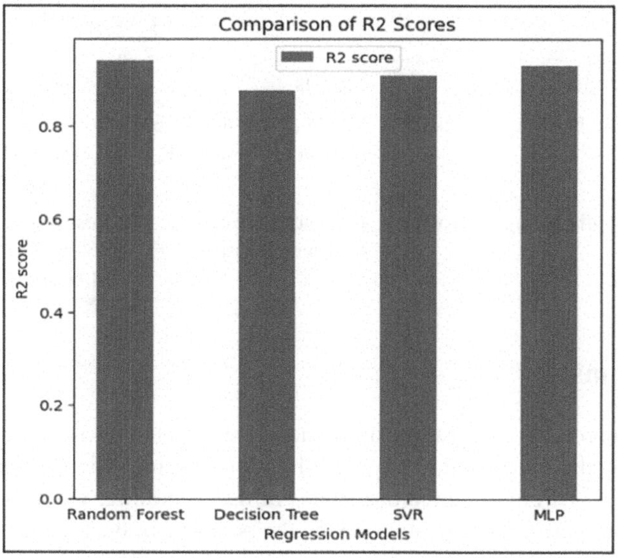

Fig. 7. Comparison of R^2 Scores

5 Discussion and Comparison

In the final discussion of our work, it is essential to address related studies, especially those utilizing the same dataset. We can discuss the results obtained in these studies, as well as the outcomes of our own study using different models.

In our study, we employed four different models: MLPR, Decision Tree (DT), Random Forest (RF), and Support Vector Regression (SVR). Among these models, the random forest yielded the best results, with an RMSE performance of 0.24 and an R^2 score of 0.94. This indicates that our random forest model succeeded in predicting target values very accurately.

In comparison, a study [6] used two models, random forest and gradient boosting, to analyze the same dataset. They achieved an R^2 score of 0.92 and a Mean Absolute Error (MAE) of 0.45. While these results are also promising, they are slightly lower than those of our study.

In conclusion, our study demonstrated that the random forest model outperformed the other three models we employed. These results are comparable to other works using similar models. However, further research is crucial to refine model performances and understand the factors influencing their outcomes.

Table 3 shows a comparative between our study and the other work, based on the dataset used, the models employed, the evaluative measures utilized, and the results obtained:

Table 3. Comparison of Results with Studies Using the Same DataSet

	DataSet	Models	Results
Our Work	LengthOfStay Microsoft	MLPR, Decision Tree, Random Forest, SVR	Best result with Random Forest: RMSE: 0.24, R2 Score: 0.94*
[6]	LengthOfStay Microsoft	Random Forest, Gradient Boosting	Best result with Random Forest: R2 Score: 0.92, MAE: 0.44

6 Conclusions

This paper explores the integration of artificial intelligence (AI) and machine learning (ML) for predicting hospital stay durations, leveraging Microsoft's "LengthOfStay.CSV" dataset. Through meticulous stages of data handling, we successfully developed accurate predictive models, showcasing the potential of these advanced techniques to enhance hospital management practices. The application of AI and ML demonstrated their capacity to improve resource planning, care optimization, and cost management within healthcare institutions. Despite these promising outcomes, it's crucial to emphasize the complementary role of predictive models to clinical judgment, serving as decision support tools.

Future research endeavors can explore two key avenues. Firstly, the development of disease-specific predictive models tailored to conditions like cardiac diseases, respiratory issues, or cancers promises more nuanced and accurate predictions. This targeted approach ensures a more personalized and effective utilization of predictive modeling.

Secondly, there exists untapped potential in exploring new variables that may influence patient stay durations. While our study utilized a comprehensive dataset, further investigations into detailed medical histories, specific test results, and comorbidities could enhance the precision of predictive models, contributing to a more comprehensive understanding of the factors impacting hospital stay durations.

Acknowledgment. This research is backed by the Algerian General Direction of Scientific Research and Technological Development (DGRSDT) and the LAMIS Laboratory.

References

1. Riley, R.D., van der Windt, D., Croft, P., Moons, K.G.M. (eds.): Prognosis Research in Healthcare: Concepts, Methods, and Impact. Oxford University Press, Oxford (2019). https://doi.org/10.1093/med/9780198796619.001.0001
2. Gentimis, T., Alnaser, A.J., Durante, A., Cook, K., Steele, R.: Predicting hospital length of stay using neural networks on MIMIC III Data. In: 2017 IEEE 15th Intl Conf on Dependable, Autonomic and Secure Computing, 15th Intl Conf on Pervasive Intelligence and Computing, 3rd Intl Conf on Big Data Intelligence and Computing and Cyber Science and Technology Congress(DASC/PiCom/DataCom/CyberSciTech), pp. 1194–1201 (2017). https://doi.org/10.1109/DASC-PICom-DataCom-CyberSciTec.2017.191

3. Rahman, M.A., Honan, B., Glanville, T., Hough, P., Walker, K.: Using data mining to predict emergency department length of stay greater than 4 hours: derivation and single-site validation of a decision tree algorithm. Emerg. Med. Australas. **32**, 416–421 (2020)
4. Abd-Elrazek, M.A., Eltahawi, A.A., Abd, M.H., Abd-Elwhab, M.N.: Predicting length of stay in hospitals intensive care unit using general admission features. Ain Shams Eng. J. **12**, 3691–3702 (2021). https://doi.org/10.1016/j.asej.2021.02.018
5. Alghatani, K., Ammar, N., Rezgui, A., Shaban-Nejad, A.: Predicting intensive care unit length of stay and mortality using patient vital signs: machine learning model development and validation. JMIR Med. Inform. **9**, e21347 (2021). https://doi.org/10.2196/21347
6. Mekhaldi, R.N., Caulier, P., Chaabane, S., Chraibi, A., Piechowiak, S.: Using machine learning models to predict the length of stay in a hospital setting. In: Rocha, Á., Adeli, H., Reis, L.P., Costanzo, S., Orovic, I., Moreira, F. (eds.) Trends and Innovations in Information Systems and Technologies, pp. 202–211. Springer, Cham (2020). https://doi.org/10.1007/978-3-030-45688-7_21
7. Hachesu, P.R., Ahmadi, M., Alizadeh, S., Sadoughi, F.: Use of data mining techniques to determine and predict length of stay of cardiac patients. Healthc. Inform. Res. **19**, 121–129 (2013). https://doi.org/10.4258/hir.2013.19.2.121
8. Andersson, O.: Predicting Patient Length Of Stay at Time of Admission Using Machine Learning (2019)
9. Chuang, M.-T., Hu, Y.-H., Tsai, C.-F., Lo, C.-L., Lin, W.-C.: The identification of prolonged length of stay for surgery patients. In: 2015 IEEE International Conference on Systems, Man, and Cybernetics, pp. 3000–3003 (2015). https://doi.org/10.1109/SMC.2015.522
10. Li, J.-S., Tian, Y., Liu, Y.-F., Shu, T., Liang, M.-H.: Applying a BP neural network model to predict the length of hospital stay. In: Huang, G., Liu, X., He, J., Klawonn, F., Yao, G. (eds.) Health Information Science, pp. 18–29. Springer, Heidelberg (2013). https://doi.org/10.1007/978-3-642-37899-7_2
11. Predicting Hospital Length of Stay [online]. https://microsoft.github.io/r-server-hospital-length-of-stay/. Accessed 14 Jan 2024

Multi-label Textual Data Augmentation Using BERT Based on Transformer Model

Roumaissa Rouabhi[1], Nacer Eddine Hammami[2], Nabiha Azizi[1](✉), Nacer Eddine Benzebouchi[1], and Rim Chaib[1]

[1] LabGed Laboratory, Badji Mokhtar University, Annaba, Algeria
azizi@labged.net

[2] Computer Engineering Department, College of Engineering and Computer Science, Mustaqbal University, Buraydah, Saudi Arabia

Abstract. In recent years, the healthcare field has taken a turn towards the AI domain, with applications such as image segmentation and medical report classification leading researchers to delve deeper into this area. On the other hand, the natural language processing has taken another turn with the arrival of the transformer models, the deep learning pre-trained models on a large corpus of documents, permitting an optimal use of medical text documents. However, the lack of diversification in text data more specifically multilabel data can be a constraint to get performant results. In this study, we are utilizing BioBERT for multilabel text classification, with the aim of achieving more refined results by implementing a data augmentation approach using a BERT-based transformer model.

Keywords: AI · medical field · nlp · transformers · text augmentation · multilabel classification · BERT · BioBERT

1 Introduction

In spite of the continuous production of extensive text data on a daily basis, its full potential remains untapped due to the abundance of disorganized information and the challenges associated with its efficient processing [1, 2]. Over the recent years, the field of text mining has experienced substantial growth and expansion, capitalizing on advancements in artificial intelligence. Text mining is a process that converts unstructured data into structured data, enabling further analysis through the application of Natural Language Processing (NLP). NLP is a subdomain of Artificial Intelligence that focuses on the interaction between computers and humans through natural language [3]. NLP applications encompass a wide range of tasks, including language translation, sentiment analysis, question answering, and speech recognition [4].

M. R. Laouar et al. (Eds.): ICISAT 2023, LNNS 981, pp. 164–173, 2024.
https://doi.org/10.1007/978-3-031-60591-8_14

Over the years, various deep learning neural network models have been explored. For instance, Zhang et al. [5] introduced the Deep Convolution Neural Network, which comprises two models incorporating word embeddings and a convolution stage [4]. In another study titled "Multi-Label Text Classification using Attention-based Graph Neural Network," Kumar et al. [6] describe an innovative method for modeling label interdependencies through a graph attention network. This model makes use of both a feature matrix and a correlation matrix to capture and examine the significant relationships among labels, subsequently constructing classifiers for the assigned task.

In order to train classifier models base with efficiency, a substantial volume of data is essential [7, 8]. To address this obstacle, various data augmentation methods have been created to expand the training dataset artificially, all while avoiding additional expenses related to labeling. These techniques comprise EDA (Easy Data Augmentation), BT (Back-Translation), and SSMBA (Self-Supervised Manifold-based Data Augmentation). While these approaches have demonstrated their effectiveness, they do not consistently guarantee diversity and the preservation of label syntax.

In the upcoming study, two models which are BERT and BioBERT are investigated with a medical dataset focused on cardiovascular diseases. The use of these models is especially to achieve two objectives: data augmentation and multilabel classification. BioBERT is built on top of BERT (Bidirectional Encoder Representations from Transformers), a pre-trained language model that was specifically pre-trained on a large corpus of biomedical text.

2 Transformers

The transformers models can be presented as new family of pre-trained deep learning models were introduced in the widely. The particularity of transformer models lies in their adoption of the two parts encoder and decoder, as shown in Fig. 1 below:

It must be noticed that some variants have been pre-trained with either one or both of these components. For instance, in this study, we utilized the BERT model, which is a transformer-based just on an encoder and which is already pre-trained on large corpora.

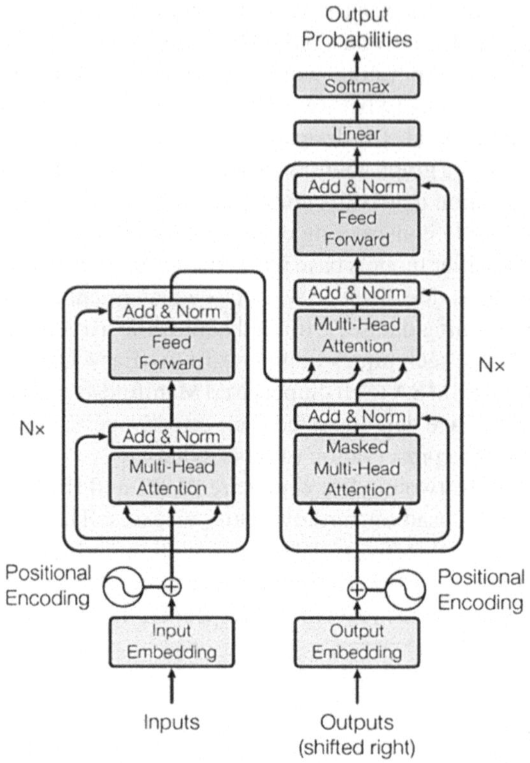

Fig. 1. The Encoder-decoder transformer Model architecture [9].

BERT (Bidirectional Encoder Representations from Transformers) was first introduced by Google researchers at the end of 2018. As mentioned in the first chapter, BERT is an autoencoder deep learning model. The core idea behind this model is the use of the Masked LM technique which is randomly masking words in a sentence then trying to predict them depending on the context of the whole sentence in a parallel, rather than sequential, manner.

In the current work we're using the BERT base uncased version for data augmentation which is a neural network-based natural language processing (NLP) model that has been pre-trained on large amounts of text data. The "base" variant of BERT is one of several model sizes available, and it has 12 layers of transformers and 110 million parameters. The "uncased" version indicates that the model was trained on lowercase text, without preserving capitalization, and does not include any special tokens or symbols for masked language modeling tasks [9]. The transformer encoder uses a self-attention mechanism to compute contextualized representations of words in a sentence (Fig. 2). It can be resumed as follow:

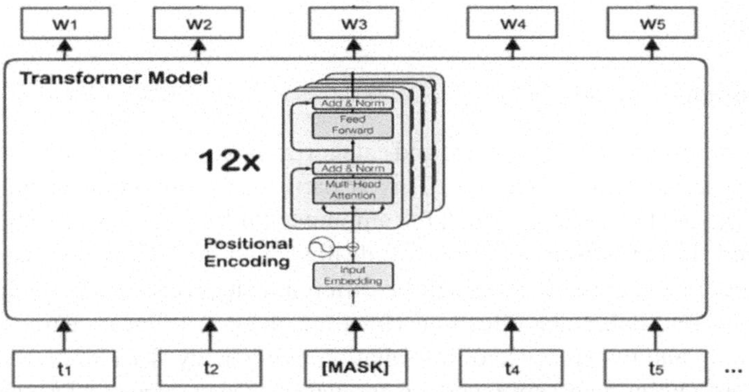

Fig. 2. The twelve encoder blocks of Bert based Transformer [10].

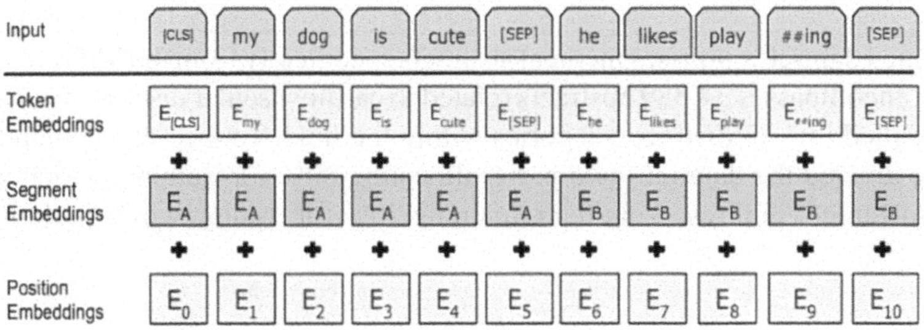

Fig. 3. BERT input representation [11].

As illustrated in the following figure (Fig. 4), the Biobert architecture closely resembles BERT but exhibits significant distinctions in its pre-training phase.

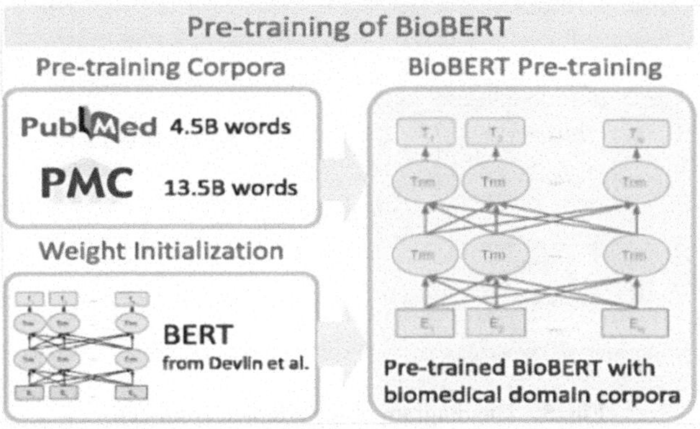

Fig. 4. Overview of the pre-training of BioBERT [8].

3 Method

3.1 Motivation

Text classification can help to organize and categorize large volumes of text data, making it easier to search, analyze, and use. This is particularly important in fields such as scientific or medical research, where large amounts of information need to be processed and organized. In the present case, we are dealing with multilabel text data, meaning that the output of the classification will be different categories for every inputted text, which leads to multilabel classification. However, the constraint in textual data is the lack of diversity and the presence of information redundancy. To address this issue, the augmentation technique was employed. Our goal is to propose a medical decision support system using multilabel text classification with a focus on addressing data diversification.

3.2 The Used Dataset

The Ohsumed dataset comprises medical abstracts specifically from the MeSH categories of 1991. It encompasses 34,389 abstracts related to cardiovascular diseases from the year 1991, classified into 23 distinct categories within the field. To meet the demands of the model, we divided the data into two parts, allocating 80% for training, which amounts to 27,511 abstracts, and reserving the remaining 20% for testing, which includes 6,878 abstracts [12].

3.3 Approach

In this study, the two models BERT and BioBERT are investigated to satify two objectives which are text augmentation and classification. As training is a crucial step for transformers, we fine-tuned our models for the augmentation and multilabel classification tasks. The functionality of our models was validated in the test phase, where we evaluated the performance on an unlabeled text test set the entire process is described in Fig. 5.

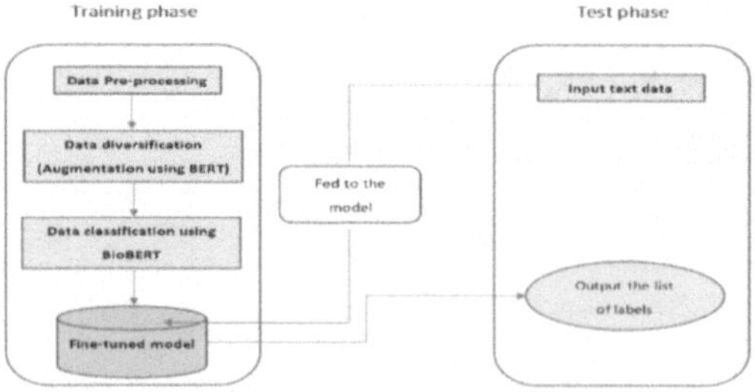

Fig. 5. The diagram of the adopted approach.

3.4 Training Phase

Data preprocessing. As the data has multiple categories, one-hot encoding is applied on the training set to represent categorical data as numerical inputs for the model. This ensures that the labels are mutually exclusive and enables the use of categorical data as input for the model that expects numerical inputs. The training data is then tokenized using the BertTokenizer tool from Bert base cased. The tokenization stage utilizes a subword tokenization method, such as byte pair encoding (BPE) or sentence piece, which dissects words into smaller subword components.

Augmentation with BERT using NLPAug. Our contribution involves injecting noise into the text in a uniform or random manner, using the transformer-based BERT model as mentioned in before and the *NLPAug* library, specifically utilizing the Contextual Word *EmbsAug* method. In certain methods to perform the augmentation, NLPAug uses pre-trained deep learning models [13].

Two different noise injection approaches are proposed in this study, each of which resulted in different performances showed in the results section. The first approach generated two samples for every input text, while the second approach varied randomly between one and three samples, detailed in Algorithm 1 and Algorithm 2, respectively.

Algorithm 1 Text augmentation with BERT model (uniform samples generation)

Require: *df*,text,labels
Ensure: *newdf*
  $onehotlabels \leftarrow$list(labels.values)
  $aug \leftarrow ContextualWordEmbsAug(modelpath =' bert - base - uncased', action = substitute)$
  $numsamples \leftarrow 2$
  for $i \leq numsamples$ **do**
    $augmentedtext \leftarrow aug(text)$
  end for
  for each augmented text **do**
    Join the words into a single string.
  end for
  $newdf \leftarrow newdf('text' : augmentedtext,' one_hot_labels' : [labels] * num_samples)$
  return *newdf*

Algorithm 2 Text augmentation with BERT model (random samples generation)

Require: df,text,labels
Ensure: $newdf$
$onehotlabels \leftarrow$list(labels.values)
$aug \leftarrow ContextualWordEmbsAug(modelpath =' bert - base - uncased', action = substitute)$
$minsamples \leftarrow 1$
$maxsamples \leftarrow 3$
$numsamples \leftarrow random(minsamples, maxsamples)$
for $i \leq numsamples$ **do**
  $augmentedtext \leftarrow aug(text)$
end for
for each augmented text **do**
  Join the words into a single string.
end for
$newdf \leftarrow newdf('text' : augmentedtext,' one_hot_labels' : [labels] * num_samples)$
return $newdf$

Data classification. The classification task involves generating a list of labels for each text sample using existing algorithms. Among these models, we focused our study on utilizing BioBERT base-cased model for analyzing the obtained results. It should be noted that the variable in this study is the choice of transformer-based models for multi-label text classification implemented by OWAIS AHMAD who used it on the PubMed Articles.[Kaggle].

Evaluation. The F1 score is adopted to evaluate proposed approaches, taking into account precision and recall measures, and the Flat validation accuracy on the other hand a simpler metric that measures the proportion of correctly classified instances in the validation set. By using both metrics, we can get a more comprehensive view of the model's performance. Regarding the loss function we selected the pytorch BCEWithLogitsLoss() funtion. It combines the BCELoss (Binary Cross Entropy Loss) and the sigmoid function into a single function. The function takes in the model's logits and the target labels, and returns the binary cross entropy loss.

4 Results

The results shown in Table 1 demonstrate the F1 validation accuracy and F1 test validation accuracy scores of our BioBERT classifier model according to the number of epochs for both augmented and non-augmented data.

Table 1. Results of the training phase.

The dataset	Type of data	Number of epochs	F1 accuracy	Test F1 accuracy
Pubmed articles	Non-augmented	6	0.87	0.85
Ohsumed	Non-augmented	12	0.76	0.73
Ohsumed	Non-augmented	20	0.77	0.74
Ohsumed	Augmented (Uniform)	20	0.88	0.86
Ohsumed	Augmented (Uniform) combined with the original data	20	0.92	0.94
Ohsumed	Augmented(Random)	20	0.89	0.87
Ohsumed	Augmented(Random) combined with the original data	20	0.96	0.94

We added the results of the Pubmed articles to show the improvements we have made as the Pubmed articles were only trained for 6 epochs with a score accuracy of 87%, according to the author it's the adviced number. We can see the approaching.

Additionally, we noticed an enhancement in accuracy following the application of BERT for augmentation stage. There was an 11% improvement in accuracy for the uniform method and 9% for the random augmentation technique. Notably, the random augmentation approach, as shown by the training loss graph in Fig. 6, exhibited the lowest training loss of 0.01.

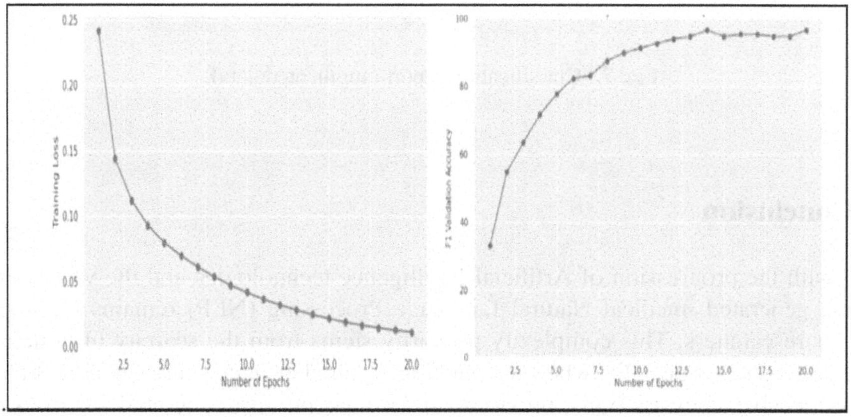

Fig. 6. Training loss and F1 score of the BioBERT model for multi-label classification.

In Fig. 7, the testing metrics for all 23 labels is presented, which include precision scores, recall scores, and F1 scores, organized by data type. Remarkably, the label

C03 consistently achieved high scores across all data types, even though it represents a minority class. Notably, when uniform augmentation was applied, label C03 achieved a remarkable precision score of 100%. These results surpass the performance of previous state-of-the-art works that used the same dataset [14, 15].

Conversely, label C22, which is also a minority class, exhibited poor performance in the non-augmented data, with a precision score of 0%. However, in the augmented data, label C22 displayed remarkable enhancement, attaining a precision score of 93%.

The application enables users to input text, which undergoes processing by the BioBERT model. Subsequently, the application generates labels based on the majority vote among the predicted labels.

Test F1 Accuracy: 0.9435959429177969
Test Flat Accuracy: 0.8646598376497874

	precision	recall	f1-score	support
C08	0.97	0.92	0.94	1561
C06	0.97	0.95	0.96	1816
C01	0.95	0.95	0.95	1512
C07	0.93	0.88	0.90	320
C09	0.96	0.92	0.94	412
C14	0.97	0.96	0.97	3632
C13	0.97	0.95	0.96	961
C22	0.87	0.82	0.85	314
C23	0.95	0.89	0.92	5774
C12	0.96	0.95	0.96	1539
C15	0.96	0.90	0.93	796
C02	0.94	0.89	0.91	672
C05	0.95	0.91	0.93	1026
C04	0.97	0.96	0.97	3800
C03	0.98	0.95	0.96	260
C10	0.94	0.93	0.94	2358
C17	0.95	0.94	0.95	983
C21	0.93	0.95	0.94	1772
C19	0.96	0.85	0.90	538
C18	0.96	0.95	0.95	1137
C20	0.94	0.97	0.95	1872
C16	0.95	0.88	0.91	663
C11	0.96	0.95	0.95	621
micro avg	0.96	0.93	0.94	34339
macro avg	0.95	0.92	0.94	34339
weighted avg	0.96	0.93	0.94	34339
samples avg	0.96	0.95	0.95	34339

Fig. 7. Classification report (augmented data).

5 Conclusion

Even with the progression of Artificial Intelligence technologies and the vast volume of data generated, medical Natural Language Processing (NLP) remains a complex task for researchers. This complexity primarily stems from the scarcity of structured and varied data, especially when it comes to multilabel data, which is essential for comprehensive analysis. While previous studies have put forth potential solutions like data augmentation, retaining the data's context while ensuring accurate labeling has proven to be a formidable challenge.

The proposed study deals with data augmentation model that takes into account context preservation and data diversity, followed by a multilabel classification model that achieved promising results, and finally. As expected given their track record, the implemented BERT and BioBERT models delivered exceptional results in NLP tasks, boasting an impressive F1 score of 96%. For future perspective, we will work on improving our augmentation model for imbalanced data therefore erasing any minor classes.

References

1. Hazel, K., Daecheol, W., Seong, J.O., Jeong-Won, C., Yo-Sub, H.: ALP: Data Augmentation using Lexicalized PCFGs for Few-Shot Text Classification (2021)
2. Benzebouchi, N.E., Azizi, N., Aldwairi, M., Farah, N.: Multi-classifier system for authorship verification task using word embeddings. In: 2018 2nd International Conference on Natural Language and Speech Processing (ICNLSP), pp. 1–6 (2018)
3. Bayer, M., Kaufhold, M.A., Reuter, C: A Survey on Data Augmentation for Text Classification (2021)
4. Bohra, A., Nguyen, H., Khatwani, D.: AutoWS: Automated Weak Supervision Framework for Text Classification (2023)
5. Kostovska, A., Doerr, C., Dzeroski, S.: Explainable Model-specific Algorithm Selection for Multi-Label Classification (2022)
6. Gargiulo, F., Silvestri, S., Ciampi, M.: Deep Convolution Neural Network for Extreme Multi-label Text Classification (2018)
7. Pal, A., Murugan, S., Sankarasubbu, M.: Multi-Label Text Classification using Attention-based Graph Neural Network (2020)
8. Devlin, J., Chang, M.-W., Lee, K., Toutanova, K.: BERT: Pre-training of Deep Bidirectional Transformers for Language Understanding (2018)
9. Lee, J., et al.: BioBERT: a pre-trained biomedical language representation model for biomedical text mining. Bioinformatics (Oxford, England) (2019)
10. Vaswani, A., et al.: Attention is All you Need (2017)
11. Khalid, U., Beg, M., Arshad, M.: RUBERT: A Bilingual Roman Urdu BERT Using Cross Lingual Transfer Learning (2021)
12. Lee, J., et al.: BioBERT: a pre-trained biomedical language representation model for biomedical text mining. Bioinformatics (Oxford, England) (2020)
13. Shim, H., Luca, S., Lowet, D., Vanrumste, B.: Data augmentation and semi-supervised learning for deep neural networks-based text classifier (2020)
14. Chaib, R., Azizi, N., Hammami, N.E., Gasmi, I., Schwab, D., Chaib, A.: GL-LSTM model for multi-label text classification of cardiovascular disease reports. In: 2022 2nd International Conference on Innovative Research in Applied Science, Engineering and Technology (IRASET), pp. 1–6 (2022)
15. Chaib, R., Azizi, N., Zemmal, N., Schwab, D., Belhaouari, S.B.: Improved multi-label medical text classification using features cooperation. In: Saeed, F., Mohammed, F., Al-Nahari, A. (eds.) Innovative Systems for Intelligent Health Informatics (IRICT 2020). LNDECT, vol. 72, pp. 61–71. Springer, Cham (2021). https://doi.org/10.1007/978-3-030-70713-2_7

Big Data Analytics: Energy Forecasting Computational Intelligence Methods

Seemant Tiwari(✉)

Department of Electrical Engineering, Southern Taiwan University of Science and Technology, No. 1, Nantai Street, Yongkang District, Tainan City 71005, Taiwan
da82b203@stust.edu.tw

Abstract. With forecasting models growing more common, energy forecasting will be utilized to enhance the energy infrastructure's layout, management, and application. Energy is one of the most significant accelerators for social and environmental improvement and improving economies. To consistently and successfully fulfill consumer electrical requirements, it is necessary to use efficient methods, make cost-effective deliveries, and adhere to a timeframe. Estimating the generation of electricity, particularly from providers of clean energy, and consumer load is essential because power plants are so reliant on the unpredictable behavior of the environment. The reliability of predictions must be increased if we are to speed up the judging process. Although big data can handle huge-scale details and detect relationships provided to deep learning methods which decrease mistakes over traditional approaches, big data had only recently started to be employed in energy forecasting. The initial phase in enhancing the dependability of clean energy production and modernizing the overall grid is to investigate artificial intelligence and machine learning technologies in this endeavor and to determine their benefits and drawbacks. A forecasting model was developed using each of the data separation approaches studied in this study. There are now technical problems with energy forecasting algorithms that need to be rectified. The current moment and forthcoming years are the timeframes for forecasting in the short term. The medium-term projection window includes the upcoming days to weeks. Long-term predictions are made in terms of either months or years. These problems were noted, and solutions were also suggested. In our view, big data is crucial for forecasts' accuracy.

Keywords: Big data · Data transformation and extracting features · Predictions for renewable energy · Electricity demand prediction

1 Introduction

Seemingly all aspects of life for people have been changed by technological advances, especially the methods used by companies to promote what they offer to clients. In addition to the now-familiar improvements, the latest developments in both marketing and technology include a multidisciplinary strategy [1], including the World Wide Web, more computer power, cell phones, and software, and further extreme breakthroughs.

M. R. Laouar et al. (Eds.): ICISAT 2023, LNNS 981, pp. 174–190, 2024.
https://doi.org/10.1007/978-3-031-60591-8_15

Digital technology and the development of technology for telecommunication and information are quickening the shift to intelligent energy infrastructure, wherein data play an amazing yet underutilized function [2]. For using artificial intelligence (AI) approaches that result in multiple big data uses for energy systems, this data, which was gathered from various sources, must first be pre-processed [3]. Instead of depending on an established formula as a model, machine learning procedures utilize computer techniques for "learning" information by analyzing data.

Among the most important catalysts for the advancement of economic, ecological, and superiority of life is electricity. It takes efficient energy-efficiency techniques, economical delivery, and scheduling for the best operating to reliably and efficiently meet customer demand for electricity. Because electrical systems are so reliant on the erratic characteristics of the climate, it is crucial to forecast the production of electricity, particularly through sustainable energy sources, and customer requests for electricity. Power network administration and upkeep are inherently intricate choice-making procedures, that are made much more difficult by the growing use of sources of clean energy, which introduces additional fluctuation and unpredictability into the functioning of the power network [4]. Using conventional evaluation methods to tackle the functioning, upkeep, and development of electricity networks can be time-consuming and not necessarily result in an acceptable remedy. In this aspect, artificial intelligence strategies can assist with the operation, upkeep, and planning of electrical systems by handling enormous amounts of data, obtaining worth from them, and coping with their range and speed via far quicker computing [5]. Every facet of the electrical structure, comprising transmission, supply, production, and usage, can benefit from the application of AI.

In this research paper, Sect. 2 presents the related works on utilizing computational intelligence and big data analysis. Section 3 presents resources for data on energy. Section 4 presents an approach for obtaining features and data preparation. Section 5 discusses the technology for big data in supply networks. Section 6 presents the use of big data to predict energy. Section 7 explains making predictions with big data in other energy fields. Section 8 describes possibilities and difficulties.

2 Utilizing Computational Intelligence, Big Data Analysis

Machine learning (ML) techniques provide a way for modeling information relationships and trends to find connections and estimate future outcomes according to unexpected occurrences. Supervised learning, unsupervised learning, and reinforcement learning are all components of machine learning techniques. Therefore, ML strategies are additionally classified as regression procedures, grouping strategies, estimation of density methodologies, and dimensionality reduction strategies. Decision tree learning, rule-based association learning, artificial neural networks, deep learning support vector machines, grouping, and Bayesian networks are a few non-exhaustive instances of these techniques.

A subset of ML techniques known as computation intelligence (CI) uses techniques that mimic the way humans procedure information and thinking capabilities to interpret complicated and unpredictable bases of information. Methods for CI are a group of nature-inspired computing techniques and methods created for tackling actual based-on-data challenges where conventional mathematical and modeling approaches are inadequate because of the processes' complicated nature, unpredictability, and stochasticity. Neural networks, fuzzy logic, and evolutionary algorithms make up the fundamental CI techniques that were created to address this expanding category of problems in reality. Fuzzy logic is a well-known strategy for dealing with ambiguous and imperfect data [6]. For modeling stochastic structures, Evolutionary Algorithms rely on the technique of organic choice, as do techniques like genetic algorithms, programming with genes, and swarm intelligent optimization techniques [7]. Last, but not least, ANNs are built on mimicking the simultaneous processing and data organization arrangement of neurons in both human and animal minds, and they allow extracting features and training from experience information [8].

In Big Data Analytics, it is becoming increasingly necessary to precisely classify key data characteristics that influence the outputs, in addition to ascertaining the spatial co-families among the input factors at a particular moment in time and the correlated or temporal co-families among the input variables that shift over time. Precise projections about the way an apparatus should act during typical operating circumstances can be produced by employing appropriate analysis to find trends using these sources of information, which also enables the identification of anomalies. As a result of their effectiveness in dealing with assignments like detecting speech and identifying objects by removing various kinds of illustrations from a variety of inputs from the senses and indications, deep learning methodologies are attracting greater focus [9]. The foundation of deep learning methods is the utilization of ANNs with numerous hidden layers when training begins unsupervised to produce more advanced visualizations of tactile information that are subsequently utilized for training classification methodologies utilizing conventional supervised learning methods [10]. Convolutional neural networks offer sophisticated designs that incorporate structural components like regional reception fields and discuss the weights and combine which seek to replicate the computation of both easy and difficult neurons in the cortex. Function techniques of learning depend on supervised methodologies like deep neural networks, convolutional neural networks, and unsupervised algorithms like LSTM [11].

The perspective of deep learning as a modeling strategy along with a way to extract associations in data is underlined. The investigators contend that Big Data and deep learning have an opportunity to deliver an entirely new set of simulations and modeling implementations according to a detailed evaluation of existing implementations [12]. The study attempts to handle a very contemporary problem brought on by the enormous quantity of information that may be gathered via handheld gadgets. The group proposes an extensible learning structure for Apache Spark and investigates deep learning as a method for mobility Big Data Analyses. This could be observed by the experimental findings, the group's architecture significantly accelerates the learning procedure for deep learning algorithms that contain a lot of secret layers and variables [13]. To be able to use Big Data for road traffic forecast reasons, investigators deploy a method of deep learning that takes into consideration both temporal and spatial connections, producing excellent results [14].

Big Data Analytics has difficulties with the type of data that has been collected by many, maybe noisy resources. Such information is filled with distortion and aberrant objects, and they are vulnerable to a great deal of ambiguity. Fuzzy algorithms have demonstrated their capacity to handle this problem, producing solid structures that operate as well as even better than other, advanced methods yet still maintaining an acceptable compromise among precision in categorization and computational efficiency. Equipment using fuzzy logic, along with particular adaptable fuzzy logic structures, have shown outstanding promise for modeling and accounting for variations in people and historical context alongside an extremely manageable computing weight, providing excellent options for developing customized and user-focused structures [15]. In [16], fuzzy logic has been utilized as a basis for performing big data analytics in different application areas. A process for creating fuzzy division algorithms programmatically. Developed a method for cardiac arrhythmia identification by combining empirical information, information from specialists, and big openly accessible information collections. Their findings show a great equilibrium between precision and comprehension [17]. Numerous applications that leverage Big Data taken from online communities for the research of public sentiment have also employed fuzzy logic [18].

Another CI methodology that can satisfy the needs and difficulties of Big Data Analytics is evolutionarily algorithmic (EA). Due to EA's excellent search missions in space skills and the fact that Big Data Analysis requires topics with an extremely high level of complexity and sparsity, Big Data Analysis is a perfect fit for them [19]. In recent years, researchers have utilized EA together using signal inputs like EEG data for challenging and contemporary fields of application like multi-brain computation [20]. A remarkable outcome when viewed alongside other contemporary methods. Additionally, evolutionary algorithms have been successfully used in Big Data Analysis to solve a variety of machine learning-associated issues, including grouping also others [21].

Combining CI approaches allows for the extraction of knowledge and significance from information, providing combined approaches that may be utilized across a wide range of usage areas. These methods ought to have been tailored to both on-site computer hardware and software data handling and management needs, which may then be additionally tuned to field-reliant on limitations and movements. As a result, such techniques may be applied to create actual structures for intelligent data evaluation and guidance in choices that are versatile and suitable for a range of business and engineering uses.

3 Resources for Data on Energy

During the following many years, it is anticipated that the amount of data produced will increase [4]. Functional and not operational information constitute the two main categories into which the data providers are divided. Any measurement instruments that gather electricity and energy data, such as voltage, current, and network condition indications, are included in the functional data for the electrical system. Contrarily, non-operational information, including meteorological data, information from the power market, and established variables provided by consumers, offer important data that is vital for enabling the efficiency of electrical products. The either past or present information can be requested through the Big Data dissemination grid facilities. Following data collection, harmonizing the information is the next phase to increase its value and give a uniform framework irrespective of its measurement provider. Information preservation, commonly referred to as the information reservoir, receives and stores standardized data.

Functional data measurement devices are intelligent metering, internet connections, and data management tools that are all part of the sophisticated measuring facilities allowing communication in both directions with energy providers and clients [22]. Observing and controlling from afar are made possible by Supervisory Control and Data Acquisition (SCADA), which gathers data continuously through transmission network elements.

Non-operational data are: Irradiance, temperatures, and the velocity of the wind are examples of time-varying climate variables that are included in weather information. For power system prediction methods, for example, meteorological data is essential. Power industry data: The outcomes of comparing the hourly and instantaneous trades provide useful data including the day-ahead power cost and the quantity of power by generating technologies kind.

4 Approaches for Obtaining Features and Data Preparation

To be able to increase the predictability, effectiveness, and precision of the method for prediction in addition to its efficacy, data preparation, choosing features, and methods for extracting features are crucial. Data purification, standardization, training, dividing information, distortion detection, management of absent values, and replacement of the same with empty values are some examples of statistical measurements. According to the information being collected, one can choose from a variety of sophisticated data decomposition approaches and use the proper data deconstruction approach. With big data categorization, unstructured data may be turned into intelligent data. Reliability and precision will be improved by acquiring an uncontaminated sample; adaptability and adaptability are important in information processing [23]. To improve the projected efficiency, dual decomposed methods for preparation and approaches for fixing faults are applied [24].

It includes primarily two methods for preparing data: statistical methods and data decomposition procedures. We perform data clean-up, eliminating noise, data splitting, and accusations of misplaced value for statistical approaches. Wavelet decomposition, variation feature decomposition, and various other techniques are used as decomposition methodologies.

5 Technologies for Big Data in Supply Networks

The classifications that follow can be used to categorize power services: functioning of the electricity grid, which is in charge of making sure the supply network operates properly, managing the adaptability markets is flexible administration, and organizing, in charge of the best ways to capitalize that support long-term transmission network. The primary goal of integrating AI techniques into Big Data electrical services is to encourage and speed up the transition of the present electrical grid to an intelligent grid that is dependable, affordable, and fair to the surroundings.

5.1 Operation and Supervision

The operations and observation division is in charge of enhancing the transmission grid's efficiency and visibility very instantly. This area includes services that utilize data like topology, visibility, and finding faults. By supplying crucial data that aids in understanding the grid state and locating potential delays, SCADA and sophisticated metering systems boost LV grid management. Monitoring the operational state of the electricity network in almost actual time to determine its current condition presents a difficulty.

The goal of the topology function is to fully retrieve the power configuration of the whole LV system. The communication network has actual time-measuring equipment at every point for functional and safety reasons to guarantee a dependable, sturdy, and precise topology. Electricity system recognition and observation [4]. The lack of actual time measurement and breakage condition equipment, nevertheless, makes topologies at transmission layers frequently unreachable and impairs their observation [25]. Table 1 shows methods based on data applied to projects for distributing network topology. Table 2 lists the data providers used to estimate the distribution network topology.

Table 1. Methods based on data applied to projects for distributing network topology.

Reference	Data-Driven Methodologies
[26]	Correlation
[27]	Linear Regression
[28]	Multi-Layer Perceptron

Table 2. Lists the data providers used to estimate the distribution network topology.

Reference	Data Providers	Measurements of Input
[27]	Meters	Voltage Levels
[29]	SCADA	Current, Energy Inoculations, and Switches Condition

5.2 Defect Finding

Throughout the electrical distribution system, the defect recognition system seeks to identify and identify aberrant electrical currents. There are two distinct ways to defect identification. On the other hand, methods based on data look for patterns in measuring data collected by devices installed at various locations throughout the system that lead to failure. On the opposite hand, methods based on modeling contrast the outcomes of forecasting algorithms against actual sensor information. An electrical problem could be indicated by large residue discrepancies [30]. Data-driven methods for locating defects in the distribution network region are shown in Table 3.

Table 3. Data-driven methods for locating defects in the distribution network region.

Reference	Apparatus	Information that's Utilized to Find Errors	Data-Driven Methodologies
[31]	Micro-grid	Indicators for both voltage and current are present at both line ends	Principal Component Analysis, K-Nearest Neighbor
[32]	Electric line	Within the PLC communication group, lines impedance, reflecting factor, and route function of transfer	Multi-Layer Perceptron
[33]	Transmission grid	circuit supply detected the current	Support Vector Machine
[34]	Electrical system	Bus voltage	Convolutional Neural Network

5.3 Preventing Future Problems

To assist in supporting decision-making on the planning of repairs to foresee an impending failure, the preventative care solution aims at extracting the possibly pertinent details gathered from acquired information from sensors placed in the electrical machinery inside the supply system [35]. Preventive repair methods make regular upkeep cheaper than restoring the following failure. Significant-power windmills are the focus of the great bulk of preventative care studies in the area of power systems because they aim to lower their substantial costs of operation and upkeep. Table 4 lists the methods based on data for forecast upkeep in the context of distributed grids.

Table 4. Lists the methods based on data for forecast upkeep in the context of distributed grids.

Reference	Apparatus	Application of data to Projecting Maintenance	Data-Driven Methodologies
[36]	Hidden cables for electricity	Current and Voltage	Large and Short Term Memory
[37]	wires, breakers, converters, and circuits	Data on failures of components during system outages	Correlation Investigation
[38]	Power Transformer	Specifications, load, position, and weather information for transformers	Random Forest

5.4 Identification of Measurement Errors

To guarantee the accuracy of data and accessibility, the measurement correction program analyzes, observes, and resolves potential deviations, mistakes, or absences of entries across sources of information. An adjustment can be carried out based on the kind of abnormality found. AI techniques are an effective tool for providing missing data with predicted values. Graph-based ANN is used in conjunction with a Multi-layer Perceptron group and a Long Short Term Memory (LSTM) driven approach for bi-direction data reduction to accomplish data that is missing restoration [39]. The primary benefit of employing AI and Big Data analytics is the way it simplifies and enhances the mistake identification procedure of the continuously growing amount of data collected concerning electricity.

Operational and non-operating information can be differentiated by the following kinds of irregularities: Empty data is whenever a recorded parameter receives no information, fundamental mistakes: occur throughout data transport or assessment, and throughout data gathering, duplicate information often happens as a result of interpersonal network issues.

6 Using Big Data to Predict Energy

The administration of energy from renewable sources, sustainability matters, disorder-founded tracking, productivity, and expense administration are now problems that the electricity industry is dealing with. The use of big data analytics opens up several new prospects for intelligent energy administration. Evaluating the massive amount of actual time information and taking into account efficient variables from data collected from sensors, significantly contributes to tackling the difficulties facing the power industry [40]. Predicting solar besides wind energy, and predicting the request for electricity in the upcoming, availability, etc. are all possible uses for big data study. To extract the useful and necessary knowledge from power and meteorological data, information combination, or the combining of data from many different places could be utilized. Technologies for smart grids may employ big data to track, evaluate, store, integrate, manipulate, and anticipate power in more intelligent ways [42].

Big data and artificial intelligence (AI) could be used together to address challenging issues in intelligent energy administration and prediction. Effectiveness could be increased by using a hybrid approach that utilizes physical and statistical techniques [41]. By applying artificial intelligence or algorithmic deep learning utilizing major platforms such as Spark, Hadoop, Strom, etc., management and processing, it is possible to estimate how much energy is consumed from huge data. For the arenas of intelligent grids and the power of the web, the forthcoming version of AI is crucial. ML techniques for intelligent grid organizing, administration, functioning, and monitoring. Statistics analysis covers past green energy statistics as well as current meteorological information gathered by electricity grid monitors. Strategies for reducing disturbance and outliers, extracting features, algorithm assortment, and data conversion procedures are all part of the data preparation procedure. Enabling deep learning forecasting algorithms to become more accurate and, at last, to be integrated for automated tracking, parameter tuning and evaluation is crucial [42]. Regarding better projections, information on sources of renewable energy as well as other weather-related variables like the velocity and direction of the wind must be taken into account. Deep learning may successfully anticipate the quantity of electricity that needs to be delivered to the network by extracting substantial information from meteorological data, such as sunlight and wind power. It is difficult to accurately represent ambiguity in physical methods. Information should be evaluated in actual time by measuring the correlation between time and geographic variables to predict wind and solar energy [43].

A total of three types of predicting horizons—short-term, medium-term, and long-term—are used [44]. Table 5 classifies prediction horizons.

Table 5. Classifies prediction horizons.

Use	Prediction Type	Period
Optimum control of the electricity system, Technologies for actual time energy administration	1 h to 6 h	Short period
Maintaining a schedule	1 week to 6 months	Medium period
Decision-making aid for purchasing market commodity futures	Up to 1 year	Long period

6.1 Predicting Wind Speed and Energy

To determine the farm's capacity for the production of electricity from the wind, it is required to forecast the climate. It is essential to take into account environmental variables like temperature, pressure, and humidity as inputs to the wind energy forecast algorithm in addition to wind velocity and course. By gathering characteristics and climate parameters concerning time and providing them as input to the projection approach, predicting wind power through the use of an effective deep convolutional neural network increases the precision of power forecasts [44]. Wind energy benefits greatly from wind speed forecasts. Many models for predicting the production of wind energy in this paper, the author provides a useful model for predicting wind power [44].

6.2 Predicting Solar Energy

The accuracy of the predictions of machine learning is excellent [45]. To efficiently handle the cohort of solar energy and to reduce electrical output variations, projection knowledge is essential. Rain and temperature are just a couple of the factors that might affect sunlight. Various resources are used for contemporary climate prediction, including measurements of the skies, satellite information, photovoltaic information, and weather information. The amount of funding required differs based on the chronological resolution. Beyond such, snowy conditions, illumination, winds, sky, haze, hailstones, and sandstorms are other meteorological conditions that have an impact on the effectiveness of solar panels [46]. Superb solar energy projection presentation is achieved via mathematical averages utilizing an iterative algorithm [47]. Extreme weather occurrences can be correctly predicted by deep learning techniques. Mixed deep learning could be utilized for modeling multivariate time series information for effective solar energy prediction.

7 Making Predictions with Big Data in Other Energy Fields

To cope handle risk and ambiguity control in the transmission system, the projection system uses AI approaches for request, generations, energy cost, and adaptability projection [42]. The objective of this subsection is to highlight fresh research that uses AI approaches across the prediction area rather than to undertake a comprehensive prediction assessment.

7.1 Prediction of Production

The goal of the production projection services is to estimate the transmission network's ability to produce energy from renewable sources. Greater accuracy in projections of renewable energy networks is required due to the continued expansion of bases of clean energy and customer adaptability initiatives for electricity networks. Concerning current research that assesses AI techniques used in renewable energy systems forecasting algorithms. Several bases of green energy, taking into account varied prediction horizons, such as dispersed wind, solar, and geothermal electricity [48].

For the creation of wind and solar energy, many ML techniques including ANNs, SVMs, and Gaussian Process Regression are investigated [49]. Evaluation of ANN used in wind power infrastructure, incorporating key techniques used in predicting algorithms and highlighting advantages and disadvantages [50]. The initial data set is split up into various frequencies using the Wavelet transformation. It is used to lessen information surges and volatility [51]. Table 6 displays the prediction of production models.

Table 6. Displays the prediction of production models.

Reference	Prediction Horizon	Area	Data-Driven Methodologies
[52]	Actual-time	Australia	LSTM
[53]	Short-time	Korea (south)	MLP, LSTM
[54]	Short-time	USA	CNN
[51]	Actual-time and Short-time	China	CNN
[70]	Short-time	Taiwan	ANN
[55]	Short-time	China	LSTM

7.2 Predicting Demands

Only a single or a group of consumer habits of ingesting are predicted by a demands projection program. The growing number of customers, or consumers who both generate and use electricity, has increased the volatility and unpredictability of the customer outline because of market reply initiatives and environmental factors that, ultimately, have an impact on the customer ingesting by oneself. Estimating consumption gets much more difficult because the network's administrator is unaware of the consumption that goes on beneath every intelligent meter.

Regression approaches are particularly effective for long-time situations that require projections, while machine learning (ML) techniques dominate for forecasting with short-term horizons [56]. In [42], the author discusses methods and predictions of load. For the sake of system functioning, short-term forecasting is the main objective of numerous investigations. Long-term investigations, however, are comparatively rare. Found that long-term forecasting systems unite the domains of energy, economics, and ecological sustainability to forecast the coming years of energy in an environmentally friendly manner [57]. Table 7 displays predicting demands.

Table 7. Displays predicting demands.

Reference	Prediction Horizon	Area	Data-Driven Methodologies
[58]	Mid-time	UK	SVM
[59]	Mid-time	USA	MLP
[60]	Actual-time	China	LSTM
[61]	Actual-time	Australia	LSTM
[62]	Long-time	Cyprus	MLP, SVM

7.3 Projecting the Cost of Power

The product seeks to forecast the cost of power, which is crucial for reducing the number of businesses that acquire and sell power. The most current and important studies concerning Big Data and AI techniques for predicting the cost of electricity are carried out in [63]. In addition to SVM regression, however, tree-based theories, and deep learning theories, estimation of the cost of electricity is a common task in the research community. To overcome influenced by time difficulties with learning, RNN is being suggested. Based on LSTM and Gated Recurrent Units (GRU) in specific, time-varying pricing prediction performed exceptionally well [64].

Table 8 presents an overview of the most current artificial intelligence (AI) methods used for the prediction of energy prices in the research literature for various prediction horizons and areas.

Table 8. Presents an overview of the most current artificial intelligence methods to display predicting demands.

Reference	Prediction Horizon	Area	Data-Driven Methodologies
[65]	Short-time	USA	Neuro-Fuzzy ANN
[66]	Short-time	Canada	ELM
[67]	Short-time	Canada	LSTM, DNN, SVM
[68]	Mid-time	Belgium	LSTM
[69]	Actual-time	USA	LSTM, CNN

8 Possibilities and Difficulties

Companies in the power industry are becoming progressively more absorbed in leveraging data science and AI to address their ongoing problems. The majority of the associated big data-driven projects continue to be currently in their infancy, and big data methodologies deployed to the power sector are currently in the beginning stages of growth. This opens up fresh possibilities regarding this important and rising field of investigation. The accessibility of large volumes of information through intelligent meters and the electronic nature of the transmission network are one of the main causes for that rising attention. Intelligent meters offer details about the functioning of the network at all levels even if they were initially primarily utilized for payment. This data offers a further understanding of whether the structure functions when integrated using other networks connected by the digitalization of the transmission network or additional outside data sources. Problems posed by big data in managing energy include a number of them.

8.1 Data Gathering and Management

The accessibility and affordability of excellent collections of data are major obstacles to the development of AI methods. The amount and quality of information accessible in the power industry are not necessarily adequate to create solutions that are capable of handling intricate situations. Additionally, the reliability, uniformity, precision, and availability of information for electricity AI applications must be increased. Effective storage and information administration techniques could successfully handle such difficulties as technological advances advance.

8.2 Facilities and Engineering

To handle the collection and preservation of Big Data, organizations were required to reinforce the information and communication technology substructure through their internal networks. It might come with new detectors, more data analysis or transfer abilities, and better transfer and retention capabilities [5]. Current information may be utilized to create novel requests, however, it might be a lot more accessible if greater energy-connected activities. Data are available and as close to the actual time as possible. It is necessary to transition to a new breed of intelligent meters featuring almost immediate access to information so that advanced metering infrastructure information can be used by more operational instruments alongside those for future planning.

8.3 Analysis and Processing of Data

ML, statistical analysis, information administration, and information presentation novel analysis of data technologies are being used, the utilization of Big Data has become a lot simpler because of ongoing and lately increasing innovations, not just in the field of electricity usage. These novel procedures for analyzing information present fresh chances to address problems and launch fresh companies in this industry.

8.4 Legal, Safety, and Concerns Regarding Privacy

According to a current spike in accidents, the computerization of the power grid has made hacking an urgent concern. Throughout the Big Data chain of custody, safety, and confidentiality must also be certain to safeguard customers, as well as to reduce hazards and potential impacts on the safety of supplies.

8.5 Abilities and Experts

To exploit Big Data technology and expand on their data skills, the power industry needs qualified and well-educated workers. Such might be done by hiring professionals about others, higher-growth industries, such as economics or advertising, or by giving Ph.D. graduates a good foundation in certain energy-associated Big Data and AI technologies.

References

1. Tiwari, S., Dubey, S.: Actively encouraging multidisciplinary innovation in academia. Curr. Adv. Multidiscip. Res. 188–194 (2022)
2. Rehmani, M.H., Reisslein, M., Rachedi, A., Erol-Kantarci, M., Radenkovic, M.: Integrating renewable energy resources into the smart grid: recent developments in information and communication technologies. IEEE Trans. Industr. Inform. **14**(7), 2814–2825 (2018)
3. Alquthami, T., et al.: Analytics framework for optimal smart meters data processing. Electr. Eng. **102**, 1241–1251 (2020)
4. Front matter. Big Data Application in Power System, pp. i–ii (2018). https://doi.org/10.1016/B978-0-12-811968-6.09985-X
5. Zhou, K., Fu, C., Yang, S.: Big data driven smart energy management: from big data to big insights. Renew. Sustain. Energy Rev. **56**, 215–225 (2016)
6. Wu, D.: Fuzzy sets and systems in building closed-loop affective computing systems for human-computer interaction: advances and new research directions. In: IEEE International Conference on Fuzzy Systems, pp. 1–8 (2012)
7. Whitley, D.: An overview of evolutionary algorithms: practical issues and common pitfalls. Inf. Softw. Technol. **43**(14), 817–831 (2001)
8. Haykin, S.: Neural Networks and Learning Machines, vol. 3. Pearson Education, India (2009)
9. Del Campo, I., Finker, R., Martinez, M.V., Echanobe, J., Doctor, F.: A real-time driver identification system based on artificial neural networks and cepstral analysis. In: International Joint Conference on Neural Networks, pp. 1848–1855 (2014)

10. Hinton, G.E., Salakhutdinov, R.R.: Reducing the data dimensionality of data with neural networks. Science **313**(5786), 504–507 (2006)
11. Korekado, K., et al.: A convolutional neural network VLSI for image recognition using merged/mixed analog-digital architecture. In: Palade, V., Howlett, R.J., Jain, L. (eds.) Knowledge-Based Intelligent Information and Engineering Systems (KES 2003). LNCS, vol. 2774, pp. 169–176. Springer, Heidelberg (2003). https://doi.org/10.1007/978-3-540-45226-3_24
12. Tolk, A.: The next generation of modeling & simulation: integrating big data and deep learning. In: International Conference on Summer Computer Simulation, pp. 1–8 (2015)
13. Alsheikh, M.A., Niyato, D., Lin, S., Tan, H.P., Han, Z.: Mobile big data analytics using deep learning and apache spark. IEEE Netw. **30**(3), 22–29 (2016)
14. Lv, Y., Duan, Y., Kang, W., Li, Z., Wang, F.Y.: Traffic flow prediction with big data: a deep learning approach. IEEE Trans. Intell. Transp. Syst. **16**(2), 865–873 (2015)
15. Karyotis, C., Doctor, F., Iqbal, R., James, A.: An intelligent framework for monitoring students affective trajectories using adaptive fuzzy systems. In: IEEE International Conference on Fuzzy Systems, pp. 1–8 (2015)
16. Tiwari, S.: Industries application of type-2 fuzzy logic. In: Laouar, M.R., Balas, V.E., Lejdel, B., Eom, S., Boudia, M.A. (eds.) 12th International Conference on Information Systems and Advanced Technologies "ICISAT 2022" (ICISAT 2022). LNNS, vol. 624, pp. 167–176. Springer, Cham (2023). https://doi.org/10.1007/978-3-031-25344-7_16
17. Behadada, O., Trovati, M., Chikh, M.A., Bessis, N.: Big data-based extraction of fuzzy partition rules for heart arrhythmia detection: a semi-automated approach. Concurr. Comput. Pract. Exp. **28**(2), 360–373 (2016)
18. Bing, L., Chan, K.C.: A fuzzy logic approach for opinion mining on large scale twitter data. In: IEEE/ACM 7th International Conference on Utility and Cloud Computing, pp. 652–657 (2014)
19. Bhattacharya, M., Islam, R., Abawajy, J.: Evolutionary optimization: a big data perspective. J. Netw. Comput. Appl. **59**, 416–426 (2016)
20. Kattan, A., Doctor, F., Arif, M.: Two brains guided interactive evolution. In: IEEE International Conference on Systems, Man, and Cybernetics, pp. 3203–3208 (2016)
21. Lin, K.C., Zhang, K.Y., Huang, Y.H., Hung, J.C., Yen, N.: Feature selection based on an improved cat swarm optimization algorithm for big data classification. J. Supercomput. **72**, 3210–3221 (2016)
22. U. S. Department of Energy: Advanced metering infraestructure and customer systems, results from the smart grid investment grant program, pp. 1–98 (2016)
23. García-Gil, D., Luengo, J., García, S., Herrera, F.: Enabling smart data: noise filtering in big data classification. Inf. Sci. **479**, 135–152 (2019)
24. Dufek, A.S., Augusto, D.A., Dias, P.L., Barbosa, H.J.: Data driven symbolic ensemble models for wind speed forecasting through evolutionary algorithms. Appl. Soft Comput. **87**, 105976 (2020)
25. Arghandeh, R., Gahr, M., von Meier, A., Cavraro, G., Ruh, M., Andersson, G.: Topology detection in microgrids with micro-synchrophasors. In: IEEE Power and Energy Society General Meeting, pp. 1–5 (2015)
26. Zhang, M., Luan, W., Guo, S., Wang, P.: Topology identification method of distribution network based on smart meter measurements. In: China International Conference on Electricity Distribution, pp. 372–376 (2018)
27. Lisowski, M., Masnicki, R., Mindykowski, J.: PLC-enabled low voltage distribution network topology monitoring. IEEE Trans. Smart Grid **10**(6), 6436–6448 (2019)
28. Zhao, Y., Chen, J., Poor, H.V.: Learning to infer: a new variational inference approach for power grid topology identification. In: IEEE Statistical Signal Processing Workshop, pp. 1–5 (2016)

29. Lin, J., et al.: Situation awareness of active distribution network: roadmap, technologies, and bottlenecks. CSEE J. Power Energy Syst. **2**(3), 35–42 (2016)
30. Nithya, M., Nagarajan, S., Navaseelan, P.: Fault detection of wind turbine system using neural networks. In: IEEE Technological Innovations in ICT for Agriculture and Rural Development, pp. 103–108 (2018)
31. Cepeda, C., et al.: Intelligent fault detection system for microgrids. Energies **13**(5), 1223 (2020)
32. Letizia, N.A., Tonello, A.M.: Supervised fault detection in energy grids measuring electrical quantities in the plc band. In: IEEE International Symposium on Power Line Communications and its Applications, pp. 1–5 (2020)
33. Fei, W., Moses, P.: Fault current tracing and identification via machine learning considering distributed energy resources in distribution networks. Energies **12**(22), 4333 (2019)
34. Li, W., Deka, D., Chertkov, M., Wang, M.: Real-time faulted line localization and pmu placement in power systems through convolutional neural networks. IEEE Trans. Power Syst. **34**(6), 4640–4651 (2019)
35. Gigoni, L., Betti, A., Tucci, M., Crisostomi, E.: A scalable predictive maintenance model for detecting wind turbine component failures based on SCADA data. In: IEEE Power and Energy Society General Meeting, pp. 1–5 (2019)
36. Skydt, M.R., Bang, M., Shaker, H.R.: A probabilistic sequence classification approach for early fault prediction in distribution grids using long short-term memory neural networks. Measurement **170**, 108691 (2021)
37. Adoghe, A.U., Awosope, C.O.A., Ekeh, J.C.: Asset maintenance planning in electric power distribution network using statistical analysis of outage data. Int. J. Electr. Power Energy Syst. **47**, 424–435 (2013)
38. Kabir, F., Foggo, B., Yu, N.: Data driven predictive maintenance of distribution transformers. In: China International Conference on Electricity Distribution, pp. 312–316, 2018
39. Spinelli, I., Scardapane, S., Uncini, A.: Missing data imputation with adversarially trained graph convolutional networks. Neural Netw. **129**, 249–260 (2020)
40. Diamantoulakis, P.D., Kapinas, V.M., Karagiannidis, G.K.: Big data analytics for dynamic energy management in smart grids. Big Data Res. **2**(3), 94–101 (2015)
41. Rahman, M.N., Esmailpour, A., Zhao, J.: Machine learning with big data an efficient electricity generation forecasting system. Big Data Res. **5**, 9–15 (2016)
42. Tiwari, S.: A survey on big data analytics for load prediction in smart grids. In: Triwiyanto, T., Rizal, A., Caesarendra, W. (eds.) Proceeding of the 3rd International Conference on Electronics, Biomedical Engineering, and Health Informatics. LNEE, vol. 1008, pp. 23–44. Springer, Singapore (2023). https://doi.org/10.1007/978-981-99-0248-4_3
43. Mujeeb, S., Alghamdi, T.A., Ullah, S., Fatima, A., Javaid, N., Saba, T.: Exploiting deep learning for wind power forecasting based on big data analytics. Appl. Sci. **9**(20), 4417 (2019)
44. Tiwari, S., Ling, J.M.: A review of wind energy forecasting techniques. In: 2021 International Conference on Technology and Policy in Energy and Electric Power (ICT-PEP), pp. 213–218 (2021)
45. Tiwari, S.: Wind speed forecasting methods for wind energy generation. In: 2022 1st International Conference on Informatics (ICI), pp. 143–147 (2022)
46. Ghazi, S., Ip, K.: The effect of weather conditions on the efficiency of PV panels in the southeast of UK. Renew. Energy **69**, 50–59 (2014)
47. Liu, L., Zhan, M., Bai, Y.: A recursive ensemble model for forecasting the power output of photovoltaic systems. Sol. Energy **189**, 291–298 (2019)
48. Ahmad, T., Zhang, H., Yan, B.: A review on renewable energy and electricity requirement forecasting models for smart grid and buildings. Sustain. Cities Soc. **55**, 102052 (2020)

49. Sharifzadeh, M., Sikinioti-Lock, A., Shah, N.: Machine-learning methods for integrated renewable power generation: a comparative study of artificial neural networks, support vector regression, and gaussian process regression. Renew. Sustain. Energy Rev. **108**, 513–538 (2019)
50. Marugán, A.P., Márquez, F.P.G., Perez, J.M.P., Ruiz-Hernández, D.: A survey of artificial neural network in wind energy systems. Appl. Energy **228**, 1822–1836 (2018)
51. Wang, H.Z., Li, G.Q., Wang, G.B., Peng, J.C., Jiang, H., Liu, Y.T.: Deep learning based ensemble approach for probabilistic wind power forecasting. Appl. Energy **188**, 56–70 (2017)
52. Chen, B., Lin, P., Lai, Y., Cheng, S., Chen, Z., Wu, L.: Very-short-term power prediction for PV power plants using a simple and effective RCC–LSTM model based on short term multivariate historical datasets. Electronics **9**(2), 289 (2020)
53. Lee, D., Kim, K.: Recurrent neural network-based hourly prediction of photovoltaic power output using meteorological information. Energies **12**(2), 215 (2019)
54. Jeong, J., Kim, H.: Multi-site photovoltaic forecasting exploiting space-time convolutional neural network. Energies **12**(23), 4490 (2019)
55. Zhou, H., Zhang, Y., Yang, L., Liu, Q., Yan, K., Du, Y.: Short-term photovoltaic power forecasting based on long short term memory neural network and attention mechanism. IEEE Access **7**, 78063–78074 (2019)
56. Kuster, C., Rezgui, Y., Mourshed, M.: Electrical load forecasting models: a critical systematic review. Sustain. Cities Soc. **35**, 257–270 (2017)
57. Suganthi, L., Samuel, A.A.: Energy models for demand forecasting - a review. Renew. Sustain. Energy Rev. **16**(2), 1223–1240 (2012)
58. Ahmad, W., et al.: Towards short term electricity load forecasting using improved support vector machine and extreme learning machine. Energies **13**(11), 2907 (2020)
59. Askari, M., Keynia, F.: Mid-term electricity load forecasting by a new composite method based on optimal learning MLP algorithm. IET Gen. Transm. Distrib. **14**(5), 845–852 (2020)
60. Tan, M., Yuan, S., Li, S., Su, Y., Li, H., He, F.: Ultra-short-term industrial power demand forecasting using LSTM based hybrid ensemble learning. IEEE Trans. Power Syst. **35**(4), 2937–2948 (2020)
61. Yang, Y., Li, W., Gulliver, T.A., Li, S.: Bayesian deep learning-based probabilistic load forecasting in smart grids. IEEE Trans. Industr. Inform. **16**(7), 4703–4713 (2019)
62. Solyali, D.: A comparative analysis of machine learning approaches for short-long-term electricity load forecasting in Cyprus. Sustainability **12**(9), 3612 (2020)
63. Lago, J., Ridder, F.D., Schutter, B.D.: Forecasting spot electricity prices: deep learning approaches and empirical comparison of traditional algorithms. Appl. Energy **221**, 386–405 (2018)
64. Ugurlu, U., Oksuz, I., Tas, O.: Electricity price forecasting using recurrent neural networks. Energies **11**(5), 1255 (2018)
65. Alshejari, A., Kodogiannis, V.S., Leonidis, S.: Development of neurofuzzy architectures for electricity price forecasting. Energies **13**(5), 1209 (2020)
66. Bisoi, R., Dash, P.K., Das, P.P.: Short-term electricity price forecasting and classification in smart grids using optimized multikernel extreme learning machine. Neural Comput. Appl. **32**(5), 1457–1480 (2018)
67. Jahangir, H., et al.: A novel electricity price forecasting approach based on dimension reduction strategy and rough artificial neural networks. IEEE Trans. Industr. Inform. **16**(4), 2369–2381 (2020)
68. Khalid, R., Javaid, N., Al-Zahrani, F.A., Aurangzeb, K., Qazi, E.U.H., Ashfaq, T.: Electricity load and price forecasting using Jaya-Long Short Term Memory (JLSTM) in smart grids. Entropy **22**(1), 10 (2020)

69. Kuo, P.H., Huang, C.J.: An electricity price forecasting model by hybrid structured deep neural networks. Sustainability **10**(4), 1280 (2018)
70. Tiwari, S., Ling, J-M.: Wind speed forecasting using narx networks models. In: International Symposium on Novel and Sustainable Technology (2022)

Efficient Chronic Kidney Disease Prediction: A Comparative Analysis Using Feature Selection and Machine Learning Models

Safa Boughougal[1](✉), Mohammed Ridda Laouar[1], and Abderrahim Siam[2]

[1] Laboratory of Mathematics, Informatics and Systems (LAMIS), University of Echahid Cheikh Larbi Tebessi, Tebessa, Algeria
{safa.boughougal,ridda.laouar}@univ-tebessa.dz

[2] University of Abbes Laghrour, Khenchela, Algeria

Abstract. Chronic nephropathy is considered one of the primary global causes of mortality. Blood pressure and diabetes are the main causes of this disease. Chronic kidney disease (CKD) can be prevented by early diagnosis, which can avoid other more serious complications such as end-stage renal disease (IRT). The early prediction of CKD is a very complex problem that requires highly developed and sophisticated solutions, such as artificial intelligence (AI) especially machine learning (ML) and deep learning (DL). This can also help improve healthcare for patients with chronic kidney disease by allowing for early detection of the disease. Our work aims to propose an artificial intelligence (AI) based approach to find the best classifier or algorithm for accurate and efficient prediction of chronic kidney disease. For our proposed approach to be effective, we conduct a thorough analysis and efficient prediction of chronic kidney disease. To achieve this, we employed effective and precise data preprocessing techniques. We utilized three machine learning algorithms: Random Forest (RF), AdaBoost, and Bagging, in addition to a hybrid model (Ada_Bag). For the training of these models, three feature selection methods were incorporated: LASSO, RELIEF, and Recursive Feature Elimination (RFE). Subsequently, we compared the performance of the machine learning models across these three feature selection techniques to identify the best method for each trained model, For the Lasso technique, the Random Forest (RF) model achieved a highest accuracy of 100% and using the RFE feature selection, the hybrid Ada_Bag model achieved an accuracy = 99,1% with 10 features.

Keywords: Kidney disease · Classification · prediction · Feature selection · machine learning · RFE · LASSO · RELIEF

1 Introduction

The kidney is like a machine that filters the blood of toxic products through thousands of millions of functional nephrons. Any damage to the renal vessels causes an alteration of the renal function that is the consequence of different pathologies: chronic kidney disease, and diabetic nephropathy. Chronic nephropathy is a gradually worsening condition that affects over 10% of the global population [1].

M. R. Laouar et al. (Eds.): ICISAT 2023, LNNS 981, pp. 191–206, 2024.
https://doi.org/10.1007/978-3-031-60591-8_16

Chronic renal dysfunction is a fatal condition typically found in adults., especially in diabetics or hypertensive people [2]. Diagnosis is based on laboratory tests of kidney function such as increased creatinine, blood urea, and anemia due to a lack of red blood cells [3].

Chronic kidney disease (CKD) is identified by reduced renal function and a Glomerular Filtration Rate (GFR) [4] lower than (60 ml/min per 1.73 m), markers of kidney damage, or both [2]. According to the World Health Organization (WHO), 10% of the population worldwide suffers from kidney disease [1]. The risk of this disease is increasing every year due to late diagnosis.

In this context, it becomes essential to give predictions to patients with chronic kidney disease to bring them early detection and management. The AI techniques, including machine learning and deep learning, stand out for their capacity to process vast data volumes. They can improve the efficiency, predictive power, and precision of diagnostic tools for persistent illnesses.

AI methods, notably machine learning and deep learning, provide a promising solution for the early detection and prediction of chronic diseases. These techniques are capable of analyzing large amounts of data and providing improved performance, predictability, and accuracy in medical diagnoses. Machine learning specifically involves using algorithms to create models from datasets and make predictions based on similar data [5]. To achieve more accurate predictions, it is recommended to train machine-learning models on appropriate datasets and utilize feature selection methods, such as REF, LASSO, etc., during the training and test process [6, 7]. There are still problems that can prevent the effective prediction of this disease, according to existing research, employing methods for feature selection to enhance the efficacy and precision of the models, applying various methods of machine learning and hybrid models, and a thorough analysis of the data and these algorithms.

Our research aims to find solutions to the difficulties associated with enhancing the accuracy and efficiency of CKD predictions. Using a thorough analysis of data and machine learning models, for this we utilized the dataset of the UC Irvine Machine Learning Repository for CKD [8]. The main goal of our work is to offer a comprehensive analysis and an efficient prediction mechanism for chronic kidney disease we employed effective and precise data preprocessing techniques to ensure our data's accuracy and relevance. In our analysis, we adopted three distinct machine-learning algorithms: Random Forest (RF), AdaBoost, and Bagging, complemented by a hybrid model (Ada_Bag). The integration of feature selection methods: LASSO, RELIEF, and Recursive Feature Elimination (RFE), was imperative to our study. These methods were chosen to optimize the models by eliminating redundant or irrelevant features, thus enhancing the accuracy and efficiency of predictions. We concluded our research with an exhaustive evaluation of the machine learning models across the three feature selection techniques. This comparison aimed to determine the most effective combination of model and feature selection method for our objective.

2 Related Work

The problems of effective prediction and early diagnosis of diseases that develop progressively, silently, and non-symptomatically require sophisticated solutions such as the utilization of machine learning (ML) and deep learning (DL) techniques.

To quickly predict the severity of CKD by applying statistical methods, machine learning techniques, and neural networks, J. Xiao et al. [9], established and compared nine predictive models such as Elastic Net, lasso regression, ridge regression, support vector machine, random forest, XGBoost, neural network and k-nearest neighbor out of the dataset of 551 patients. They considered different parameters such as 13 blood tests and 5 demographic characteristics for the early prediction of chronic kidney disease. The linear models including Elastic Net, lasso regression, ridge regression, and logistic regression achieved the highest level of predictive performance with an average AUC of 87% and a precision of 80%.

In their pursuit of early diagnosis for chronic kidney disease, Baidya et al. [10], presented an artificial intelligence method, they used eight machine-learning algorithms such as Extra Tree Classifier (EXT), AdaBoost (ADB), K-Nearest Neighbors (KNN), Gradient Boosting (GB), Extreme Gradient Boosting (XGB), Decision Tree (DT), Gaussian Naïve Bayes (GNB) and Random Forest (RF) on the dataset the UCI machine learning repository, they combined the eight algorithms by voting classifier method such that there are two kinds of voting soft and hard, in this work they used soft voting to make an exact prediction. Afterward, they showed that the execution time of the hybrid model is equal to 317 ms, and to evaluate the performance of each model, they used several metrics: F1 score, precision, sensitivity, accuracy, recall, and AUC score. From the results obtained, they observed that the K-Nearest Neighbors (KNN) and Extra Tree Classifier achieved better results for the prediction of CKD with an accuracy of 99%.

Pal et al. [11], developed a predictive model for the prediction of chronic kidney disease, they used three machine-learning algorithms (Logistic Regression, Decision Tree, and Support Vector Machine). They found that the best classifier is the decision tree with an accuracy of 95.92%.

To develop a model based on machine learning (ML) models to forecast chronic kidney failure; this study focuses on four features: sex, age, comorbidities, and medications. Krishnamurthy et al. [12], applied five models: Random Forest, Decision Tree,Logistic Regression, LightGBM, and CNN on the National Health Insurance Research Database (NHIRD) dataset. The CNN model achieved the best accuracy result of 89%, while the other models (Logistic Regression, Decision Tree, Random Forest, and LightGBM) achieved an accuracy of (73.6%, 73.2%, 72.5%, and 75.1%,) respectively.

Ani et al. [13], Implemented a system to improve decision-making for clinicians based on machine learning models: LDA classifier, neural network based back propagation (BPN), probability-based Naive Bayes, Random subspace classification algorithms, decision tree, and K Nearest Neighbor (KNN) to predict chronic kidney disease. After evaluating the accuracy results, the random forest model achieved the highest accuracy of 94%, while the LDA model had an accuracy of 76%.

Hassan et al. [14], developed a prediction model based on machine learning to recognizing chronic kidney disease in its initial phases. Hassan et al. [14] used Neural Network (NN), Random Forest (RF), Support Vector Machine (SVM), Random Tree

(RT), and Bagging Tree Model (BTM) with the XGBoost feature selection method, seven features were selected: hemo, sg, pcv, al, pe and pb. The NN model achieved the highest accuracy of 100% with all 25 attributes, while the SVM had an accuracy of 100% with the seven features.

Various studies have utilized artificial intelligence in their research, especially ML models for early and efficient prediction of kidney disease, by conducting in-depth data analysis, using different feature selection methods, and comparing the accuracy of multiple classifiers and hybrid models to find the best classifier for disease prediction (Table 1).

Table 1. Summary of Previous Studies Comparison

Previous studies	Dataset	Models	Feature selection methods	Accuracy	Limitation
Ani et al. [13]	UCI	LDA classifier, NN-based back propagation (BPN), Naive Bayes, RF, DT, and KNN	NO	76%,81%,78%,94%,93%, and 90%	The authors did not use feature selection methods and they did not use data preprocessing to improve accuracy
Xiao et al. [9]	551 patient's datasets	LR, Elastic Net, lasso regression, ridge regression, SVM, RF, XGBoost, NN, and KNN	NO	82%,82%,81%,81%,81%,80%,83%,80 and 74%	No feature selection method was used in this work to improve accuracy
Krishnamurthy et al. [12]	NHIRD	Logistic Regression, Decision Tree, Random Forest, LightGBM and CNN	LightGBM	73.6%,73.2%72.5%and 75.1%	The authors focused on only 4 features (age, sex, comorbidities, and medications), not on laboratory tests
Baidya et al. [10]	UCI	EXT, ADB, KNN, GB, XGB, DT, GNB, RF	NO	99%,94%,99%,98%,95%,95%,96% and 95%	The researchers did not use any feature selection methods to improve the accuracy and efficiency of the machine learning models
Hassan et al. [14]	UCI	NN, RF,SVM, RT, BTM	XGBoost	97.5%,98.75%,100%,96,25%,97.5%	The authors used only 7 features

3 Methodology

The problem of predicting CKD is a classification problem between two situations (CKD or Not-CKD).

The main objective of our study is to conduct a thorough analysis of any data for the application of machine learning classifiers and hybrid models on the dataset. First, we collected data from the UCI Machine Learning Repository [15]. After that, we proceeded to the data preprocessing step to clean the dataset and handle missing values using imputation methods. Three feature selection methods: LASSO, Relief, and RFE were used to prevent overfitting, improve accuracy, and reduce training time. I apologize for misunderstanding. The dataset is divided into two parts: 70% for the training phase and 30% for the testing phase. After that, we trained three machine learning models (Adaptive Boosting, Bagging classifier, Random Forest) and hybrid models (Ada_Bag). The performance of the models is analyzed and compared across all features of the dataset and on the features selected by the feature selection methods, so that we can choose the best model for our dataset and the best feature selection method for each algorithm. The proposed approach was described in (Fig. 1):

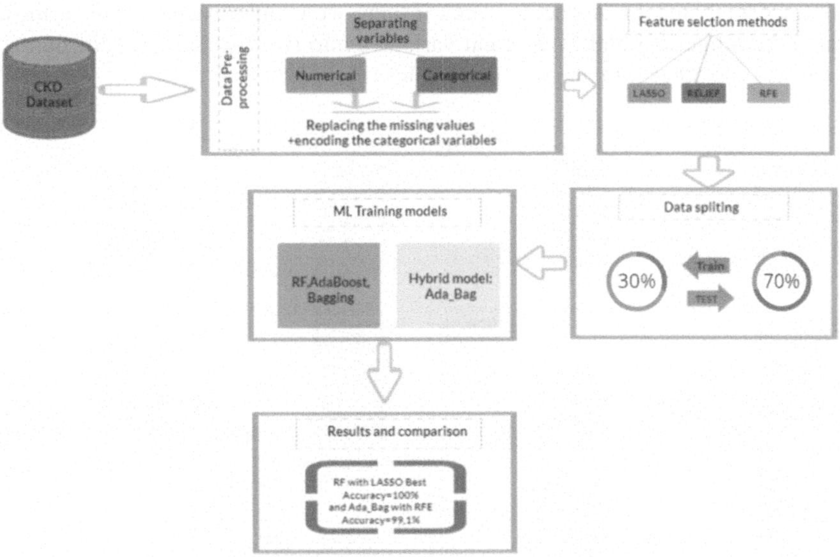

Fig. 1. Architecture of our approach.

The target class that is mentioned in the dataset that is "classification" is divided into two classes CKD = 250 and notCKD = 150, this is what the following pie chart shows in percentage too (Fig. 2):

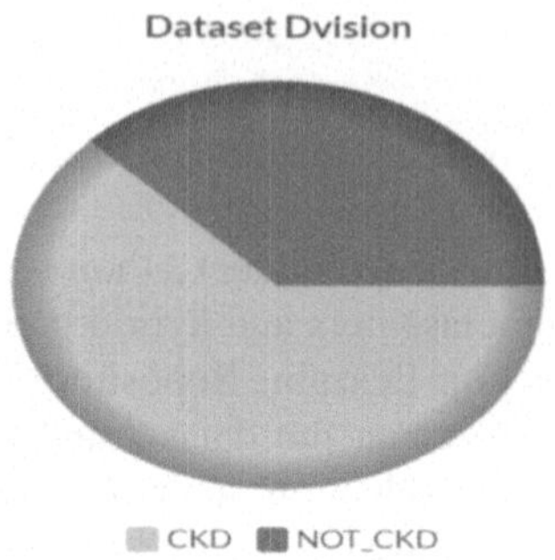

Fig. 2. Dataset division (Classification 'ckd' or 'not_ckd).

3.1 Pre-processing

Data pre-processing is considered a very important phase, as a dataset may contain redundant data, nulls and missing values, etc. Null, redundant or missing values can disturb the global learning of the model and lead to false predictions. For this reason, data pre-processing is important in order to fill in missing values, remove null values.

First, we analyzed the dataset to check if there were any missing or null values, we separated the numerical and categorical variables into two lists, one list for numerical variables and the other for categorical variables (Figs. 3 and 4).

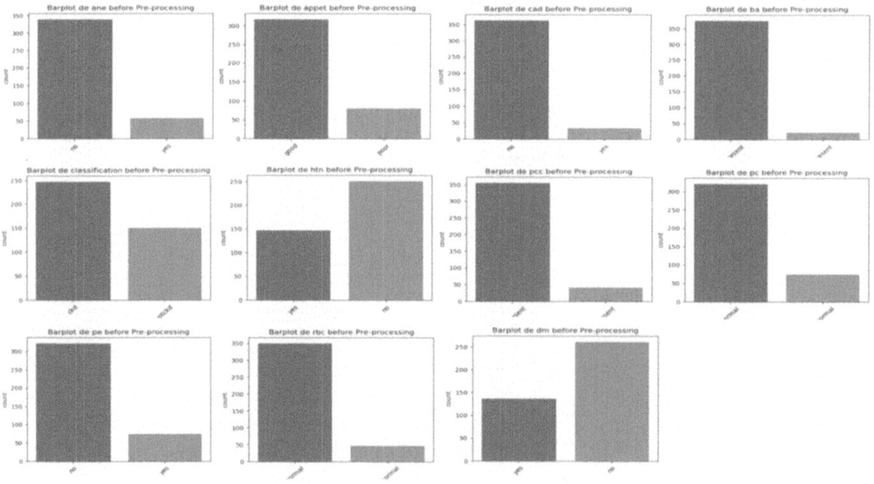

Fig. 3. Graphs represent categorical variables.

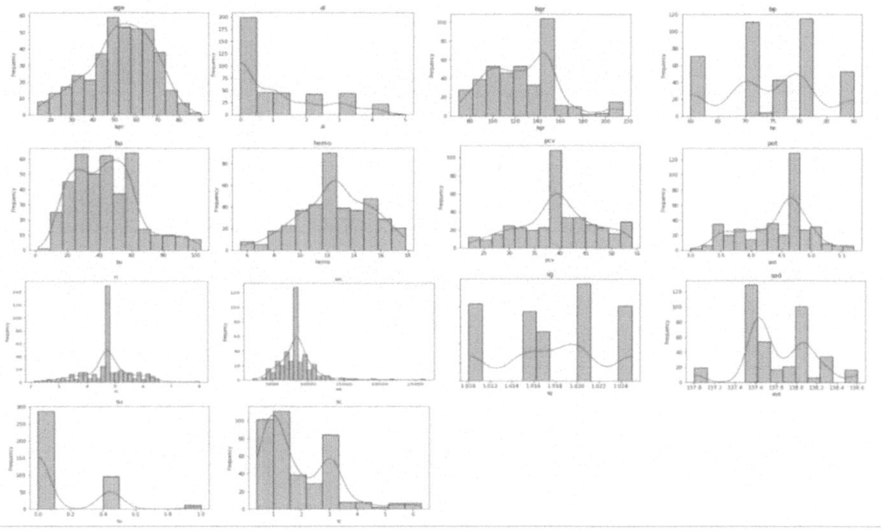

Fig. 4. Represent numerical variables.

The missing values were dealt with using imputation methods. For categorical variables used the most frequent value imputation method (mode) substituting missing values in each column with the most frequent value. For numerical variables, the "Fit" method was applied, which estimates statistics like the mean from numerical data to fill in missing values (Figs. 5 and 6).

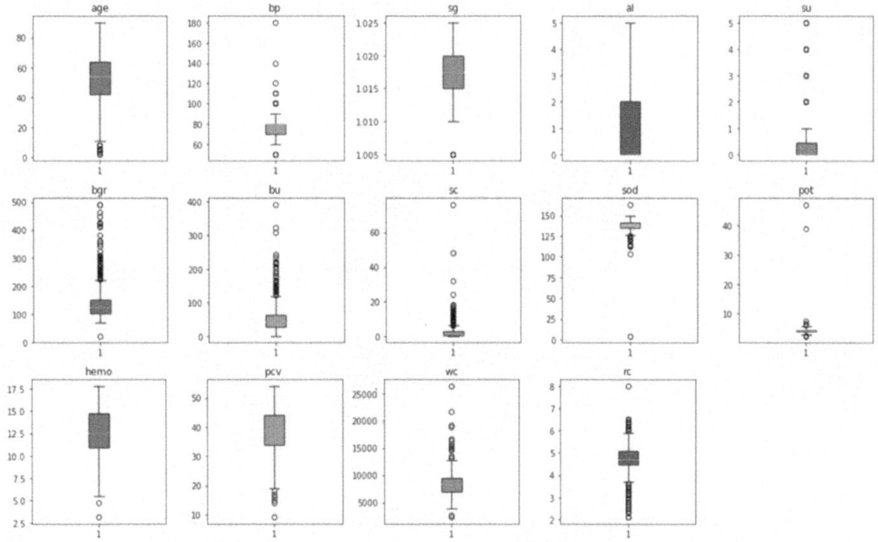

Fig. 5. Numerical variables before imputation.

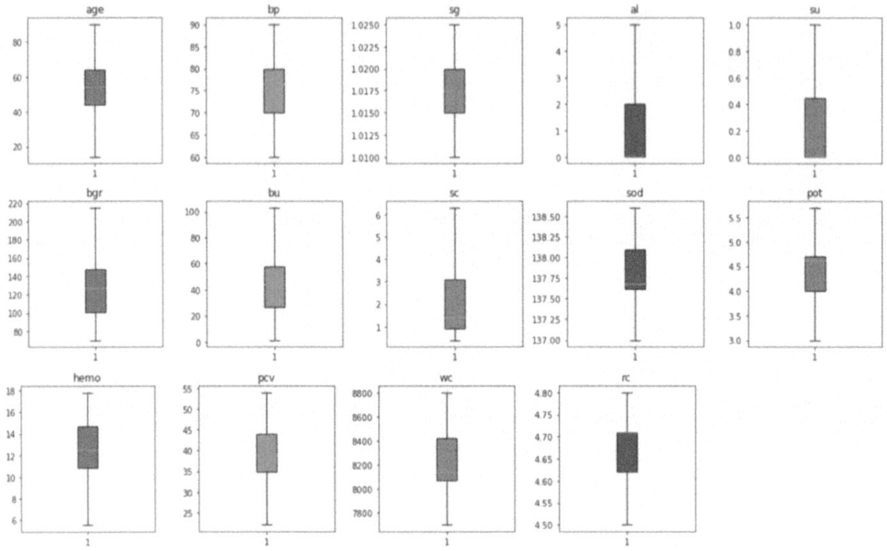

Fig. 6. Numerical variables after imputation.

Then we encoded the categorical variables into numerical variables and replaced the original data with the encoded data, which is necessary for machine learning algorithms that only accept numerical data by using the "LabelEncoder" class (Fig. 7).

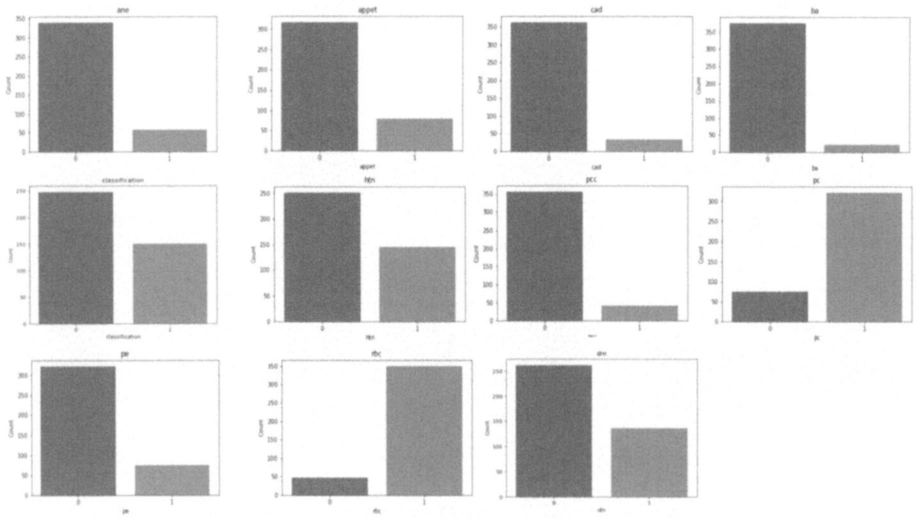

Fig. 7. Categorical variables after encoding into numerical.

3.2 Feature Selection Methods

In machine learning training, feature selection methods have a very important aspect because they improve the accuracy of machine learning models (precise prediction) and reduce the training time of these models. In our work, we have used three methods: RELIEF, Recursive feature Elimination (RFE) and Least Absolute Shrinkage and Selection Operator (LASSO).

Recursive Feature Elimination (RFE)

Recursive Feature Elimination (RFE) is a technique used in machine learning to select a subset of the most relevant features for a given dataset to improve the prediction performance of a model. By classifying a large number of features based on a specific machine learning method, RFE can optimize the prediction accuracy of the model [16, 17].

Utilizing the RFE (Recursive Feature Elimination) method for predicting chronic kidney disease (CKD), several clinical markers have surfaced as impactful. The Specific Gravity (sg), with a score of 3.25, emerged as the most significant feature, highlighting its role in determining the concentration of urine - a vital sign of kidney function. Hemoglobin (hemo), an oxygen-carrying protein in red blood cells, and Packed Cell Volume (pcv), which measures the blood volume occupied by red blood cells, also demonstrated considerable importance with scores of 1.98 and 1.82, respectively, emphasizing their role in diagnosing anemia, a common CKD symptom. On the other hand, the negative scores associated with Albumin (al), Serum Creatinine (sc), and Diabetes Mellitus (dm) at −1.72, −1.53, and −1.82 respectively indicate potential risk factors for CKD. Other features like Sodium (sod), Pus Cell (pc), and Red Blood Cells (rbc) with positive scores further highlight the multifaceted nature of CKD prediction (Fig. 8).

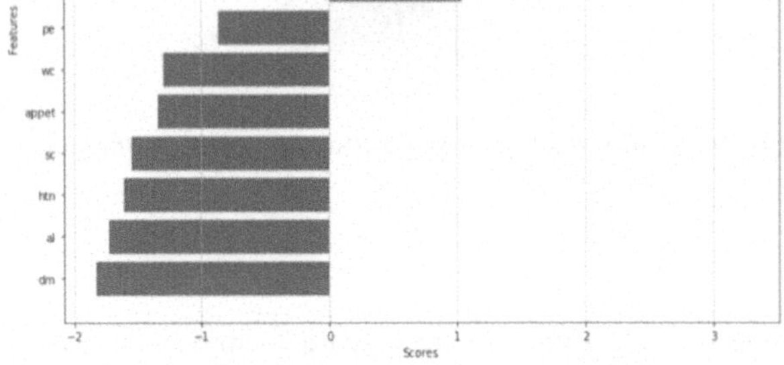

Fig. 8. The important features selected by RFE and their scores.

Least Absolute Shrinkage and Selection Operator (LASSO)

Least Absolute Shrinkage and Selection Operator (LASSO) is a feature selection method used in machine learning that reduces the coefficients of less important features to zero. Some feature coefficients become zero, and those with negative coefficients may be removed from the subset of features [18]. The method promotes important coefficients and eliminates unnecessary coefficients by re-evaluating them to zero, allowing the selection of the most important features to improve the prediction performance of the model. The Specific Gravity (sg) measures the kidney's ability to concentrate urine, with a higher 'sg' typically indicating healthy kidney function. The positive coefficient of 2.01 suggests that as 'sg' increases, the likelihood of not having CKD rises. Conversely, Albumin (al) presence in urine indicates abnormal protein leakage by the kidneys. Its negative coefficient of −1.45 suggests that higher levels of 'al' increase the risk of CKD. Similarly, a higher level of Serum Creatinine (sc) indicates impaired kidney function, and its negative coefficient of −1.01 further associates elevated 'sc' levels with a greater CKD risk. On the other hand, Hemoglobin (hemo) is a protein in red blood cells responsible for oxygen transport. The positive coefficient of 1.36 for 'hemo' indicates that higher hemoglobin levels might be linked to a lower risk of CKD (Fig. 9).

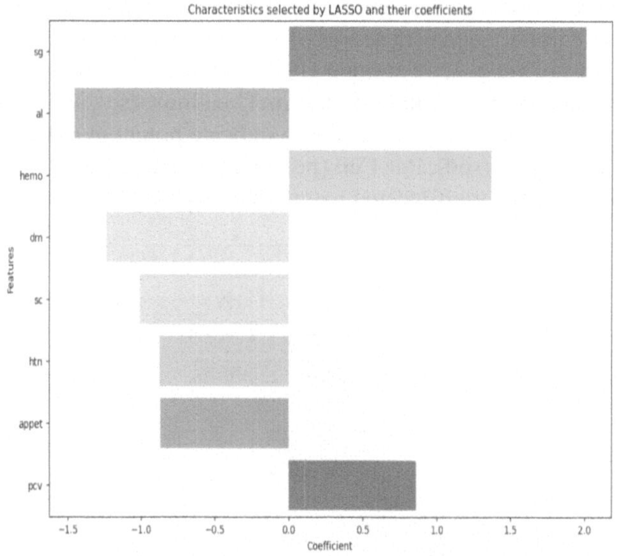

Fig. 9. Features selected by LASSO and their coefficients.

Relief

Relief is a technique for selecting important features in a dataset by assigning higher weights to them and lower weights to the remaining ones [18]. It relies on methods similar to K-nearest neighbors (KNN) to determine the weights. The Relief algorithm selects a random instance, Rj, and finds its two nearest neighbors: the nearest hit (NH) from the same class and the nearest miss (NM) from the opposite class. Based on the values of Rj, NM, and NH, it updates the relevance score W[A] for each feature A. If the difference between Rj and NH for a feature A is large, its performance value is reduced, as this is undesirable. Conversely, if the difference between Rj and NM for feature A is large, A can be used to distinguish different classes, and its weight W[A] is increased. The Relief algorithm was originally proposed by Kira and Rendell [17]. The process continues for a number of iterations, with a parameter m that can be adjusted.

Utilizing the RELIEF feature selection method for predicting chronic kidney disease (CKD), the most influential feature was Specific Gravity (sg) with a score of 0.46, highlighting its importance in indicating the concentration of substances in urine. Hemoglobin (hemo) followed with 0.39, reflecting oxygen-carrying proteins in red blood cells, and Packed Cell Volume (pcv) at 0.32, representing the blood volume occupied by red cells. Hypertension (htn) scored 0.28, emphasizing its close link to CKD, while Diabetes Mellitus (dm) stood at 0.27, reinforcing its role as a primary CKD cause. Serum Creatinine (sc) at 0.24, urine Albumin (al) at 0.16, and Pus Cell (pc) at 0.12 further underline kidney function and potential associated infections. These scores, as delineated by RELIEF, offer valuable insights into CKD's onset or progression. The following figure shows these features with their scores (Fig. 10).

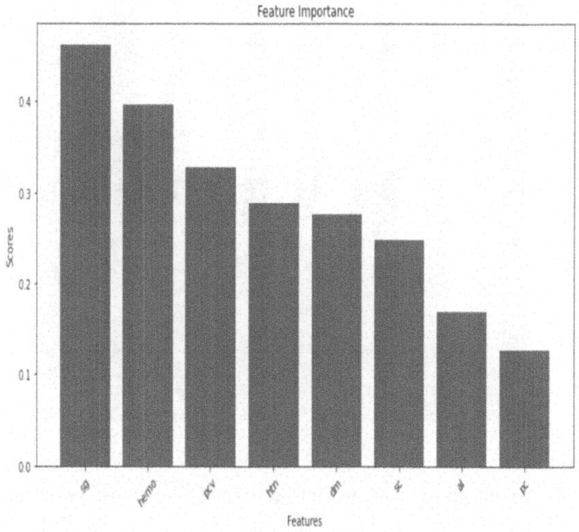

Fig. 10. Features selected by RELIEF and their scores.

4 Results and Discussion

In Machine Learning, it is important to divide the data set into two subsets, one for the training phase (train-set) and the other for the test phase (test-set). The test-set is used for the final validation of the model in order to make sure of its capacity of generalization on unknown data. As we did in this paper, 70% for the training phase and 30% for the test phase. The dataset is used for the training, testing and validation phase of the models of classification. To evaluate the performance of machine learning models and hybrid models implemented in our research, we employed metrics such as accuracy, precision, recall, and F1_score. Using the UCI machine-learning repository's dataset for chronic kidney disease (CKD), we evaluated four models: Random Forest (RF), AdaBoost, Bagging, and the hybrid Ada_Bag. Lasso feature selection identified eight critical attributes: Specific Gravity (Sg), Albumin (al), Hemoglobin (hemo), Diabetes Mellitus (dm), Serum Creatinine (sc), Hypertension (htn), Appetite (appet), and Packed Cell Volume (pcv). Among these, RF exhibited perfect metrics, suggesting potential overfitting, while AdaBoost consistently scored around 0.978 for accuracy, precision, recall, and F1-score. Bagging achieved a remarkable 0.991 accuracy and perfect precision. The hybrid Ada_Bag combined AdaBoost and Bagging strengths, securing an accuracy of 0.975 with an exceptional recall of 1.0. Though all models displayed strong performances, considerations around overfitting and the balance between precision and recall are crucial for health-related predictions (Fig. 11).

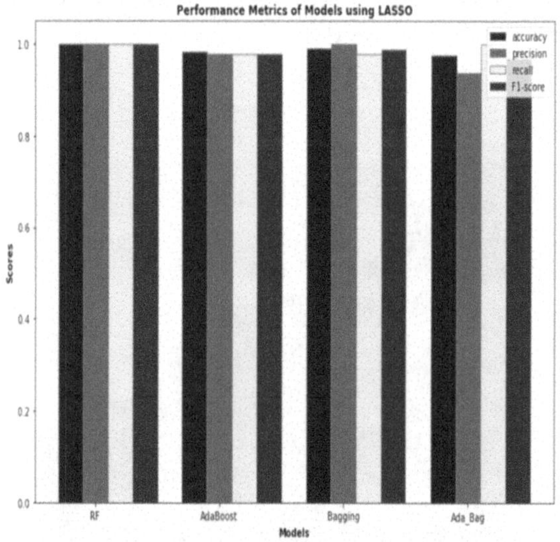

Fig. 11. Performance of models with LASSO feature selection.

Using the RELIEF feature selection method, eight important features has been selected ('sg', 'hemo', 'pcv', 'htn', 'dm', 'sc', 'al', and' pc') were identified. Among the evaluated models, Random Forest (RF) emerged superior with a striking 0.991 accuracy, supported by a 0.978 precision and a flawless recall of 1.0. AdaBoost showed promise with 0.958 accuracy and a notable 0.976 precision, yet a slightly diminished recall of 0.913. Bagging, boasting a 0.977 accuracy and impeccable precision, hinted at minor recall enhancements. Conversely, the Ada_Bag hybrid, while registering a 0.92 accuracy, struck a balance with a perfect recall, indicating its cautious yet accurate positive predictions (Fig. 12).

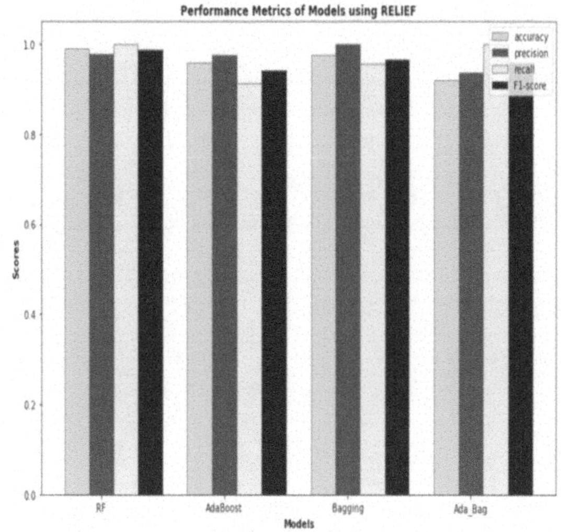

Fig. 12. Performance of models with RELIEF feature selection.

Finally, thirteen features have been selected with RFE technique such as ('sg','hemo','pcv','pc','sod','rbc','pe','wc','appet','sc','htn','al' and 'dm'), among others. In assessing the performance of various models using this refined feature set, the Random Forest (RF) model rendered a commendable accuracy of 0.99, bolstered by a perfect precision and an F1-score of 0.989, though its recall slightly trailed at 0.978. Both AdaBoost and Bagging mirrored each other in terms of accuracy (0.983 for AdaBoost and 0.991 for Bagging) and displayed impeccable precision, with the F1-score of Bagging slightly outperforming AdaBoost. However, the (Ada_Bag) hybrid stood out distinctively, achieving a stellar accuracy of 0.991, and flawless metrics across precision, recall, and F1-score, emphasizing its unparalleled capability in this particular prediction context (Fig. 13).

Comparing the Accuracy of Different Models with Feature Selection Methods
In our research on the efficient and accurate prediction of chronic kidney disease, we employed a variety of machine learning models to evaluate their performance. These models were trained using three different feature selection methods: Lasso, RELIEF, and RFE. Based on the accuracy results, for the Lasso method, the Random Forest (RF) model showed a perfect accuracy of 100%, followed closely by Bagging with 99.1%, AdaBoost with 98.5%, and the hybrid Ada_Bag model at 97.5%. When trained using the RELIEF feature selection method, RF achieved an accuracy of 99.1%, Bagging 97.7%, AdaBoost 95.8%, and the hybrid Ada_Bag model lagged at 92%. Lastly, under the RFE feature selection, both the RF and Bagging models displayed an impressive accuracy of 99.1%, with AdaBoost slightly behind at 98.3% and the hybrid Ada_Bag model matching the leaders with 99.1%. These results indicate the robustness of the RF model across different feature selection techniques and highlight the potential for hybrid models.

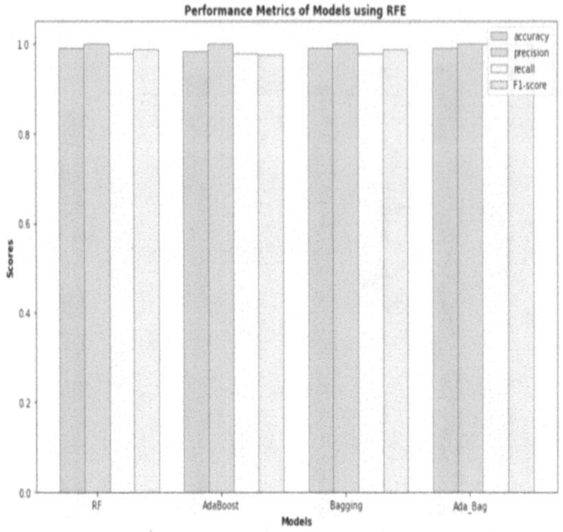

Fig. 13. Performance of models with RFE feature selection.

Comparison of Our Proposed Approach with Existing Work

In our comparison between our predicted models and existing works, our methods consistently outperformed or enhanced existing accuracy rates. For instance, while the existing Random Forest (RF) model had an accuracy of 88.78% [19], our techniques using LASSO, RELIEF, and RFE achieved 100%, 99.1%, and 99%, respectively. The AdaBoost model, previously at 96% accuracy [10], was refined by our feature selection to reach up to 98.3%. The Bagging model showed a similar trend, improving from an initial 97% [20] to 99.1% with our LASSO method. Importantly, our study introduced the hybrid Ada_Bag model, which demonstrated impressive accuracies, peaking at 99.1% with RFE. Unlike existing works, our study uniquely incorporated three feature selection methods and introduced the Ada_Bag model, underscoring our innovative approach to chronic kidney disease prediction (Table 2).

Table 2. A Comparison of Accuracy Between our approach and existing works.

Models	Accuracy of existing works	Accuracy using LASSO	Accuracy using RELIEF	Accuracy using RFE
RF	88,78% [19]	100%	99,1%	99%
AdaBoost	96% [10]	98,3%	95,8%	98,3%
Bagging	97% [21]	99,1%	97,7%	99,1%
Ada_Bag	/	97,5%	92%	99,1%

5 Conclusion

Chronic kidney disease is a serious complication of diabetes and blood pressure, it develops asymptomatically, and an early diagnosis of this disease requires very advanced techniques like machine learning and deep learning. Various studies have already attempted to predict kidney disease through of machine learning and deep learning. For an efficient and accurate prediction underscores the significance of precise data preprocessing and accurate selection of machine learning models. By incorporating algorithms like Random Forest (RF), AdaBoost, Bagging, and the hybrid Ada_Bag, we were able to delve deep into the intricacies of the disease's predictors. The integration of feature selection methods, specifically LASSO, RELIEF, and Recursive Feature Elimination (RFE), further enhanced our models by ensuring the relevance and optimization of input features. Our comparative analysis revealed key insights into the most efficient combinations of models and feature selection methods. For efficient prediction of CKD, we will follow a similar approach as the other studies. Notably, using the Lasso technique, the Random Forest (RF) model achieved the highest accuracy of 100%. Conversely, with the RFE feature selection, the hybrid Ada_Bag model reached an accuracy of 99.1% with a mere 10 features. Looking ahead to future endeavors, there is potential to elevate this research by leveraging deep learning algorithms, refining the datasets, and exploring alternative feature selection methodologies. As we progress, such research not only paves the way for enhanced diagnostic tools but also underscores the evolving synergy between healthcare and technological advancements.

References

1. Kovesdy, C.P.: Epidemiology of chronic kidney disease: an update 2022. Kidney Int. Suppl. **12**(1), 7–11 (2022). https://doi.org/10.1016/j.kisu.2021.11.003
2. Krzesinski, J.M., Scheen, A.J.: La maladie rénale diabétique: prise en charge actuelle et perspectives d'avenir. Rev. Med. Suisse **11**(483), 1534–1542 (2015)
3. Malkina, A.: Maladie rénale chronique (2022). https://www.msdmanuals.com/fr/professional/troubles-génito-urinaires/maladie-rénale-chronique/maladie-rénale-chronique
4. Borisagar, N., Barad, D., Raval, P.: Chronic kidney disease prediction using back propagation neural network algorithm, pp. 295–303 (2017). https://doi.org/10.1007/978-981-10-2750-5_31
5. Shailaja, K., Seetharamulu, B., Jabbar, M.A.: Machine learning in healthcare: a review. In: 2018 Second International Conference on Electronics, Communication and Aerospace Technology (ICECA), pp. 910–914 (2018). https://doi.org/10.1109/ICECA.2018.8474918
6. Mienye, D., Sun, Y., Wang, Z.: An improved ensemble learning approach for the prediction of heart disease risk. Inform. Med. Unlocked **20**, 100402 (2020). https://doi.org/10.1016/j.imu.2020.100402
7. Mishra, J., Tarar, S.: Chronic disease prediction using deep learning, pp. 201–211 (2020). https://doi.org/10.1007/978-981-15-6634-9_19
8. Chronic_Kidney_Disease (2015). https://archive.ics.uci.edu/dataset/336/chronic+kidney+disease
9. Xiao, J., et al.: Comparison and development of machine learning tools in the prediction of chronic kidney disease progression. J. Transl. Med. **17**(1), 1–13 (2019). https://doi.org/10.1186/s12967-019-1860-0

10. Baidya, D., Umaima, U., Islam, M.N., Shamrat, F.M.J.M., Pramanik, A., Rahman, M.S.: A deep prediction of chronic kidney disease by employing machine learning method. In: 2022 6th Int. Conf. Trends Electron. Informatics (ICOEI 2022) - Proc., no. Icoei, pp. 1305–1310 (2022). https://doi.org/10.1109/ICOEI53556.2022.9776876
11. Pal, S.: Chronic kidney disease prediction using machine learning techniques. Biomed. Mater. Devices (2022). https://doi.org/10.1007/s44174-022-00027-y
12. Krishnamurthy, S., et al.: Machine learning prediction models for chronic kidney disease using national health insurance claim data in Taiwan. Healthcare **9**(5), 546 (2021). https://doi.org/10.3390/healthcare9050546
13. Ani, R., Sasi, G., Sankar, U.R., Deepa, O.S.: Decision support system for diagnosis and prediction of chronic renal failure using random subspace classification. In: 2016 International Conference on Advances in Computing, Communications and Informatics (ICACCI), pp. 1287–1292 (2016). https://doi.org/10.1109/ICACCI.2016.7732224
14. Hassan, M.M., et al.: A comparative study, prediction and development of chronic kidney disease using machine learning on patients clinical records. Human-Centric Intell. Syst. **3**(2), 92–104 (2023). https://doi.org/10.1007/s44230-023-00017-3
15. UCI machine learning repository: Early stage of Chronic kidney disease dataset. https://archive.ics.uci.edu/ml/datasets/chronic_kidney_disease
16. Li, L., Ching, W.-K., Liu, Z.-P.: Robust biomarker screening from gene expression data by stable machine learning-recursive feature elimination methods. Comput. Biol. Chem. **100**, 107747 (2022). https://doi.org/10.1016/j.compbiolchem.2022.107747
17. Granitto, P.M., Furlanello, C., Biasioli, F., Gasperi, F.: Recursive feature elimination with random forest for PTR-MS analysis of agroindustrial products. Chemom. Intell. Lab. Syst. **83**(2), 83–90 (2006). https://doi.org/10.1016/j.chemolab.2006.01.007
18. Ghosh, P., Azam, S., Jonkman, M., Shultana, S., Beeravolu, A.R.: Efficient prediction of cardiovascular disease using machine learning algorithms with relief and LASSO feature selection techniques. **9** (2021). https://doi.org/10.1109/ACCESS.2021.3053759
19. Chittora, P., et al.: Prediction of chronic kidney disease - a machine learning perspective. IEEE Access **9**, 17312–17334 (2021). https://doi.org/10.1109/ACCESS.2021.3053763
20. Baidya, D., Umaima, U., Islam, M.N., Shamrat, F.M.J.M., Pramanik, A., Rahman, M.S.: A deep prediction of chronic kidney disease by employing machine learning method. In: 2022 6th International Conference on Trends in Electronics and Informatics (ICOEI), pp. 1305–1310 (2022). https://doi.org/10.1109/ICOEI53556.2022.9776876
21. Baidya, D., Umaima, U., Islam, M.N., Shamrat, F.M.J.M., Pramanik, A., Rahman, M.S.: A deep prediction of chronic kidney disease by employing machine learning method. In: 2022 6th International Conference on Trends in Electronics and Informatics (ICOEI), Tirunelveli, India, pp. 1305–1310 (2022). https://doi.org/10.1109/ICOEI53556.2022.9776876

Blockchain Limitations in the Safe Storage and Transfer of Electronic Health Records

Chahinez Mérièm Bentaouza(✉)

Mostaganem University, Mostaganem, Algeria
chahinez.bentaouza@univ-mosta.dz

Abstract. The purpose of this article is to identify the use of blockchain in healthcare, and specifically the possible disadvantages. The patient's medical records are crucial for personal ownership, as they are a record of their entire care history. Numerous aspects of society have been transformed by the internet, similar to how Blockchain technology has brought about significant changes in various aspects of both business and society, during the past two decades. The technology of Blockchain, a distributed database, is widely recognized for its role in keeping track of transaction details for digital currencies, like Bitcoin. Despite this, the basic mechanics for executing transactions by individuals and organizations are not up to date for the 21st century. Furthermore, we discovered potential conducting an additional literature review for application areas. By identifying all the weaknesses and potential improvements in current solutions to this subject, we were able to introduce new research concepts.

Keywords: Blockchain · Electronic medical records · Keys · Limitations · Scalability

1 Introduction

Blockchain technology is commonly used to refer to various things, and it can be difficult to distinguish between Bitcoin Blockchain, Ethereum Blockchain, or smart contracts. Distributed ledgers are being discussed, which involves a list of transactions that is duplicated on multiple computers instead of being kept on one server. The use of blockchain has been demonstrated for its usefulness in recording events like transactions.

Medical records could benefit from this, given their frequent inaccuracy and discrepancies. It is the responsibility of healthcare professionals to ensure confidentiality, medical records must be accurate, complete, and only accessible to to those who have been authorized. Information storage by healthcare providers can be difficult due to their different systems. As an example, if hackers could manipulate a patient's prescription information. In essence, the higher the quality and security of health information, the higher the quality of care offered.

M. R. Laouar et al. (Eds.): ICISAT 2023, LNNS 981, pp. 207–212, 2024.
https://doi.org/10.1007/978-3-031-60591-8_17

2 Related Work

2.1 The Impact of Blockchain on Healthcare and Medical Records

Medical data sharing is the biggest challenge facing health care systems worldwide that assure data integrity and patient privacy by working with known and unknown stakeholders for various purposes. Despite the improvement in data standards, each electronic health record (EHR) stores data through distinct workflows.

The medical fraternity is facing a challenge in creating a trusted environment for decision making due to the unclear who recorded what and when. The importance of care coordination and EHR access throughout the care continuum has raised concerns about how to ensure that patient data can be viewed, edited, and shared by multiple providers. While keeping a record of diagnoses, medications, and services rendered that is authoritative and up-to-date [2].

Data sharing logistics isn't the sole issue with the EHR problem. Patients should have the right to privacy and control over their own information. It should be the primary concerns for any solution that is worthy of an important factor to consider in a national health care system. Hospitals in the United States have been anticipating the arrival of EHRs as a new age of standardization and high-quality healthcare for nearly a decade [3].

While health care data has become more accessible thanks to federal laws and incentive programs, most hospital systems are still unable to share their data easily or safely. It is important to look for solutions that enable patients to make their own decisions about who they reveal their identity with, where they stay anonymous, and what data they want to share [2].

2.2 Builder Application

The picture above illustrates that every block contains the hash of the previous one to connect them (Fig. 1).

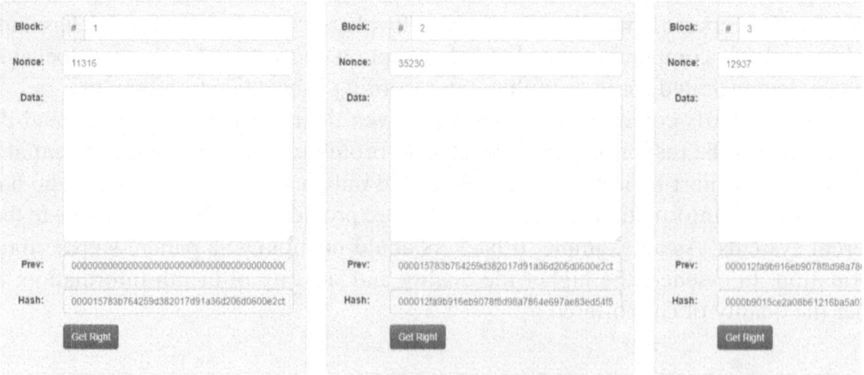

Fig. 1. Coherence of blocks

The picture above shows that every block of the blockchain has a transaction that displays the amount, so all transactions and senders are stored in blocks that create a distributed blockchain (Fig. 2).

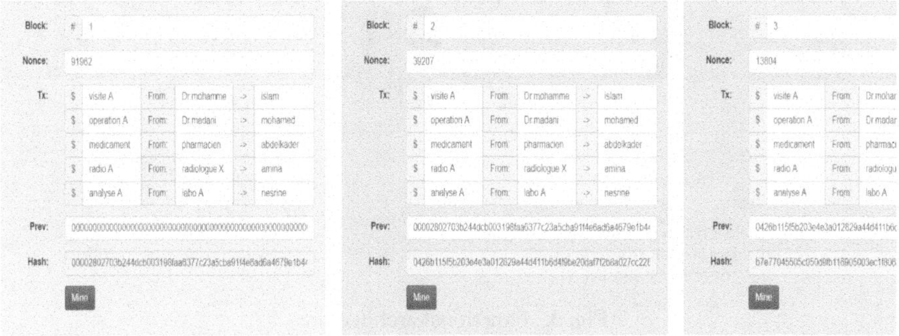

Fig. 2. Invalidation of integrity view

We shouldn't get too attached to the cloud computing analogy. Although blockchain infrastructure is similar to a layer of cloud computing infrastructure, it does not allow us to replace it. In comparison to a traditional cloud service like Amazon Web Services or Digital Ocean, Blockchain VMs may be too expensive. However, they can be advantageous for a few decentralized applications.

In the future, client nodes may be able to communicate directly with each other in situations where blockchains are too costly or unreliable.

2.3 Blockchain Model for Health Care

To be effective in healthcare, a blockchain must be publicly accessible and provide three crucial areas of technology are scalability, access security and data privacy (Fig. 3).

To confirm that the document or image is authentic, a healthcare provider will create a digital signature when producing a medical document that includes prescriptions, lab tests, pathology results, or MRIs. For storage, health data will be encrypted and sent to the data storage location.

The blockchain keeps a record of the health record and user ID every time information is saved to the data lake.

Notification is given to the patient that their blockchain has been updated with health data. Patients would be able to add healthcare data in a similar way by using digital signatures and encryption in mobile apps and wearable sensors [4] (Fig. 4).

Health data analysis can be performed by the healthcare provider using a customized best-in-class application [5].

The shared public ledger would not provide any information to an adversary due to the presence of only hashed pointers and encrypted information in the transactions [6].

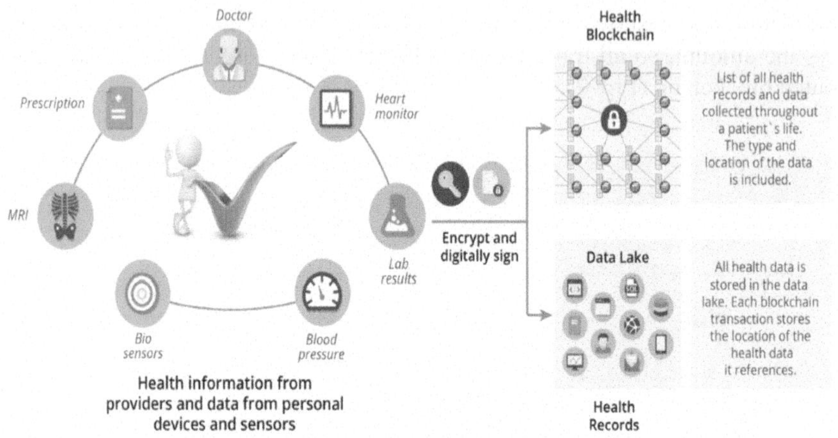

Fig. 3. Functional architecture

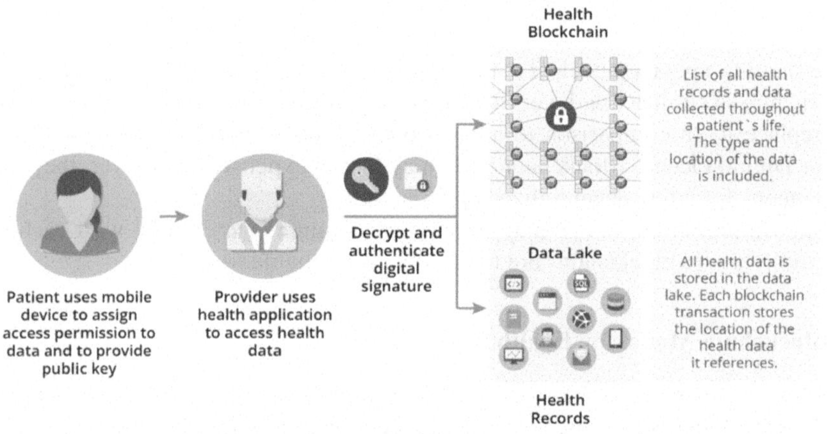

Fig. 4. Transaction architecture

3 Blockchain Limitations

The Bitcoin crypto currency was the starting point for blockchain technology [7] in 2008. This new and young technology is a problem because it has only recently started currently being utilized in different areas and is still being tested for various applications, including banking [8]. Due to the lack of academic works been done in this manner, there is a limitation [9]:

- Scalability
- Cost
- Privacy & Security
- Regulation & Compliance Risks
- Lack of Interoperability

Public blockchains are accessible to all, while private blockchains are restricted to a few players [10] (Fig. 5).

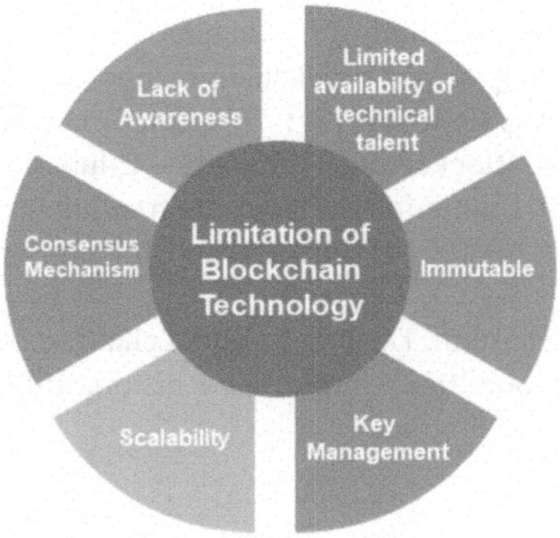

Fig. 5. Limitation of Blockchain

4 Conclusion

The first step is to present the problem, which explains the need for digitization of medication plans and how blockchain technology is the suitable technology to implement this application. Blockchain technology is outlined, given its new and unfamiliar IT structure [11]. After that, the problem has been solved by proposing a design. Demonstration was achieved by creating a system of smart contracts. It is possible to develop such an application and suggest guidelines for designing a blockchain system that meets the defined requirements [12]. The issue of privacy-handling applications is the focus of a discussion [13] about the applicability of different blockchain designs.

References

1. Attou, A., Bentaouza, C.M.: (2018). http://e-biblio.univ-mosta.dz/handle/123456789/9659
2. Coinmarketcap: Crypto-Currency Market Capitalizations (2016). https://coinmarketcap.com/. Accessed 15 Mar 2018
3. Coinmarketcap: ZCash Charts (2017). http://coinmarketcap.com/currencies/zcash/. Accessed 01 Feb 2018
4. https://www.ibm.com/blogs/blockchain/category/blockchain-in-healthcare/
5. AD: Blockchain gaat de wereld veranderen (2016). http://www.ad.nl/nieuws/blockchain-gaat-de-wereld-veranderen~a4697f98/. Accessed 22 May 2018
6. Berkeley: BlockChain Technology: Beyond Bitcoin (2015). scet.berkeley.edu/wp-content/uploads/BlockchainPaper.pdf. Accessed 01 Feb 2018
7. Deutsch, M.: The Resolution of Conflict: Constructive and Destructive Processes. Yale University Press (1973)
8. Dwivedi, Y.K.: Handbook of Research on Contemporary Theoretical Models in İnformation Systems. IGI Global (2009)

9. Rožman, N., Corn, M., Škulj, G., Berlec, T., Diaci, J., Podržaj, P.: Exploring the effects of blockchain scalability limitations on performance and user behavior in blockchain-based shared manufacturing systems: an experimental approach. Appl. Sci. **13**, 4251 (2023). https://doi.org/10.3390/app13074251
10. https://www.javatpoint.com/limitation-of-blockchain-technology
11. Sharma, S., Rosmin, P., Bhagat, A.: Blockchain technology: limitations and future possibilities (2021). https://doi.org/10.4018/978-1-7998-2414-5.ch009. https://www.researchgate.net/publication/348119914_Blockchain_Technology_Limitations_and_Future_Possibilities
12. Ammbika, V.M., Rao, D.S.: Limitations of blockchain technology with its applications. Int. J. Recent Technol. Eng. **8**(2S11) (2019). https://www.ijrte.org/wp-content/uploads/papers/v8i2S11. ISSN: 2277-3878
13. Frederic, B., Giulio, C., Sebastian, D., Jon, F.: Blockchain scalability and the fragmentation of crypto. BIS Bulletin No 56 (2022). https://www.bis.org. ISSN: 2708-0420

A Rule-Learning Approach for the Personalization of Context-Aware Multimedia Documents Adaptation Processes

Aziz Smaala[1,2](✉), Abdelkader Moudjari[3], Asma Saighi[2,4], Zakaria Laboudi[5], and Saad Harous[6]

[1] Research Laboratory on Computer Science's Complex Systems (ReLa(CS)2), University of Oum El Bouaghi, Oum El Bouaghi, Algeria
azizsmaala@univ-oeb.dz
[2] Department of Mathematics and Computer Sciences, University of Oum El Bouaghi, Oum El Bouaghi, Algeria
[3] Department of Mathematics and Computer Sciences, Ecole Normale Supérieure, University of Constantine3, Ali Mendjeli, Algeria
[4] Laboratory of Artificial Intelligence and Autonomous Things (LIAOA), University of Oum El Bouaghi, Oum El Bouaghi, Algeria
[5] Department of Networks and Telecommunications, University of Oum El Bouaghi, Oum El Bouaghi, Algeria
laboudi.zakaria@univ-oeb.dz
[6] Department of Computer Science, College of Computing and Informatics, University of Sharjah, Sharjah, UAE
harous@sharjah.ac.ae

Abstract. The execution of multimedia documents in pervasive environments may involve the adaptation of their contents according to users context information. In this regard, many adaptation approaches have been proposed. Generally, these approaches deals with context collection, representation and interpretation without taking into account its storage and analysis. This feature helps provide prior knowledge about the past decisions that users may take in the future. Thus, we propose in this paper a rule-learning approach for the personalization of adaptation rules within context-aware multimedia documents adaptation processes. The proposal uses rule based data mining classifiers, by means of the sequential covering algorithm. Our proposal is validated through scenarios implemented in a real prototype that systematically records the context values and the actions performed accordingly. These data, are then used by the rule learning algorithm in order to personalize the adaptation actions to users. The obtained results are very satisfactory and encouraging.

Keywords: context-aware pervasive systems · multimedia document adaptation · historical users data · rule learning · personalized adaptation rules

M. R. Laouar et al. (Eds.): ICISAT 2023, LNNS 981, pp. 213–223, 2024.
https://doi.org/10.1007/978-3-031-60591-8_18

1 Introduction

Ambient intelligence aims to build environments with communication and computation capabilities so as to help mobile applications sense their environment, understand the current context and adapt accordingly. Thus, this paper focuses on the adaptation of multimedia documents (i.e. documents consisting of different media objects such as text, image, audio, video, etc.) in context-aware pervasive systems so that they comply with both users profiles, and computational and environmental constraints (e.g., noise level, location, screen size, battery level, etc.). A typical example is ubiquitous learning which refers to a learning environment supported by computers, mobile devices, connected objects and wireless networks. The learning materials are composed of different learning objects, each of which may include various media contents (i.e. video, text, audio, image, slides, etc.). In such an environment, learning objects should be adapted at every time and everywhere, based on learners contexts. For instance, if the learner consults pedagogical materials through a tablet in the lab, the system should convert auditory contents into texts.

In the literature, several multimedia documents adaptation (MDA for short) processes have been proposed [1–14]. Throughout the adaptation process, the user context is first sensed. Then, the conflicts making the documents features uncompliant with the current context are identified. Finally, adaptation actions are inferred in order to provide the adapted contents, often using "IF (conditions) THEN actions" rules. Although MDA processes have yielded interesting results, their analysis showed us that they mainly target the context collection, representation and interpretation without taking into account its storage and analysis as historical data (log data). Indeed, logging of historical data helps provide prior knowledge about the past decisions taken by the users so that they can be considered in future processing. This way, the global performance of MDA processes can be improved by leveraging various advantageous tasks such as applying machine-learning techniques and building recommendation systems.

In this work, we propose a new approach that takes advantage of the historical user data (HUD) resulting from the execution of MDA processes, in order to personalize context-aware adaptation rules to users through rule learning mechanisms. The proposal uses rule based data mining classifiers which is a very well-known technique in machine learning and data mining, by means of the sequential covering algorithm. This task refers to prediction of adaptation actions using data-driven approaches (i.e. based on data analysis and interpretation) rather than explicitly defining them. More specifically, the historical data are analysed in order to discover users' behaviours leading to personalization of context-aware adaptation rules. Our proposal is validated through scenarios implemented in a real prototype that allowed us collecting historical data of users by systematically recording the context values and the actions performed accordingly. These data, are then used by the rule learning algorithm in order to personalize the adaptation actions to users. Globally, the obtained results are very satisfactory and encouraging.

The remainder of this paper is structured as follows. Section 2 provides the main basics of MDA processes and reviews some related works. Section 3 gives the details of the proposed approach for the personalization of adaptation rules of multimedia documents. Section 4 focuses on validation aspects of the proposal by discussing its implementation as well as the obtained results. The paper ends with some concluding remarks.

2 Related Works

As the context in pervasive systems changes over time, some limitations and restrictions may arise as a result of elements of multimedia document that conflict with the current context. In this regard, many adaptation approaches have already been developed, in order to provide documents that meet as much as possible the current contexts of users [1–14]. These approaches differ essentially in the manner the adaption process is carried out as well as the types of the multimedia documents treated. As mentioned in [15], context-aware systems comprise three layers: *sensing*, *thinking*, and *acting*. Following that, we go over these features in the case of MDA processes, as shown in Fig. 1.

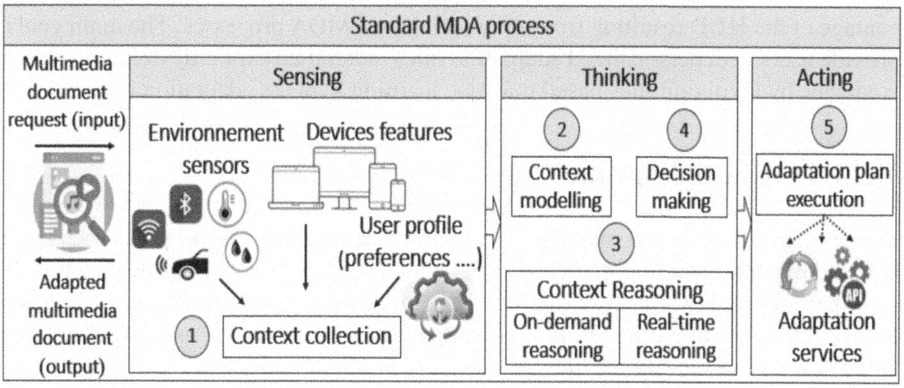

Fig. 1. The general architecture of a standard MDA process.

- **Sensing layer:** this layer is tasked with collecting the context elements values by receiving data from the physical world via various sources such as sensors (e.g. lighting, noise level, locations, network connectivity, etc.).
- **Thinking layer:** this layer tasked with processing the context values in order to take desired actions. The context information collected by the sensing layer is raw as it only consists of values measured by various sources (low-level context data). Therefore, the information gathered is first organized using an abstract context model (e.g. key-value, object-oriented, logic-based, ontology-based, etc.) to generate a high-level context data. Then, these data are used to perform a higher-level reasoning upon context information to determine the constraints that prevent playing the documents properly and thus to deduce the required actions leading to producing adapted contents, often using "IF – THEN" rules.

– **Acting layer:** the role of this layer is to execute the adaptation plan by carrying out each adaptation action, in order to fulfill the needs of users and environments. These actions are applied to the objects that make up the document so that a set of adapted contents is generated. These adapted objects are organized in a new document so that the presentation of their contents becomes more suitable to the current contextual situation.

Although MDA processes have shown interesting results [1–14], they are often built around a set of predetermined adaptation rules that do not change over time, either defined by designers or customized by users. Unfortunately, if the adaptation rules are defined by designers, these rules may partially or not at all suit users. Likewise, by allowing users to customize the adaptation rules, it is almost impossible to predict and thus cover all situations that can be encountered (i.e. rules including compound conjunctive conditions); this is because new contextual situations are continuously arising over time. Thus, it will be better if there is an efficient mechanism for the personalization of adaptation rules to users instead of defining them in a fixed way. This helps the system take automatic decisions and predictions which, in fact, is one of the main objectives targeted by ambient intelligence.

Based on this perspective, we propose through this work a new approach that takes advantage of the HUD resulting from the execution of MDA processes. The main goal is to provide users with personalized adaptation rules rather than explicitly defining them in a fixed way, by involving rule-based machine learning to make adaptation rule learning.

3 The Proposed Approach

Although there exist several methods for making data analysis, our proposal focuses on rule-based machine learning that can extract context-aware adaptation rules from HUD due to their simplicity, clarity, readability and rules base handling efficiency (adding, deleting, etc.) [16]. In particular, the sequential covering algorithm may excel in terms of reliability and versatility of rules.

3.1 Logical Structure of Log Data (Historical Data)

In order to execute rule-learning algorithms, the system should make systematic recordings of the HUD each time new data are acquired by the system. The data to be stored consists mainly in the values of the context elements and the corresponding adaptation actions resulting from changes in context over time. The logical structure of the log data takes the form of a table containing the context elements values and the corresponding actions performed by the user (see Table 1).

Table 1. The logical structure of the raw log data.

Context elements				Actions			
CE_1	CE_2	....	CE_n	Action 1	Action 2	.....	Action m
V_{11}	V_{21}	....	V_{n1}	{yes/no}	{yes/no}	{yes/no}	{yes/no}
.	.	.	.	.	.	.	.
.	.	.	.	.	.	.	.
V_{1k}	V_{2k}	....	V_{nk}	{yes/no}	{yes/no}	{yes/no}	{yes/no}

Table 2 shows a typical example for a subset of context elements values and adaptation actions performed accordingly.

Table 2. Typical example of context elements values and adaptation actions.

Context elements				Actions	
Battery	Network	Screen size	Location	Speech2Text	Text-summary
Full	Mobile data	Narrow	Public place	1	0
Low	Mobile data	Narrow	Bus	1	1
Full	WI-FI	Wide	Home	0	1

3.2 Summary of the Proposed Rule-Learning Process

The proposed learning mechanism for the personalization of context-aware adaptation rules belongs to rule-based learning. This paradigm is widely adopted in machine learning and data mining, in order to find regularities in data that can be expressed as IF-THEN rules [17–20]. Adaptation rule learning refers to the process of iteratively updating and refining the rule set based on HUD and adaptation actions performed. To do so, we need to discover the relationships between the different items in the log data, more specifically between the context elements values and the actions performed. These relationships are then represented as rules taking the form "**IF (*Antecedent*) THEN *Consequent***". The *conditions* consist in compound conjunctive conditions on context elements values while the *Consequent* consists in a set of adaptation actions. For this purpose, the process given in Fig. 2 is proposed.

Our rule-learning algorithm is based on the *sequential covering algorithm* which is used to learn a set of rules that can accurately classify instances into predefined categories. It follows a sequential process of rule generation and instance classification by iteratively refining the rule set until a satisfactory level of accuracy is achieved. In fact, although the trend is now shifting towards the use of deep learning techniques, the sequential covering algorithm and similar techniques are more suitable to our study for two main reasons. First, the rule learning algorithm should run continuously as the HUD are generated each time the context changes. Therefore, we need for an algorithm that run faster as it will periodically run. Second, we plan to adapt the rule-learning algorithm so that it runs on several devices ranging from limited computation resource to those having a computation power. Hence, we will need for an algorithm that consumes less computation resources. The sequential covering algorithm can be adapted to incorporate an adaptive learning approach for personalizing adaptation rules of multimedia documents to users. Moreover, sequential covering may excel in rule induction due to its simplicity, effectiveness and interpretability of resulting rules (i.e. interpretable rules that can provide insights into the decision-making process).

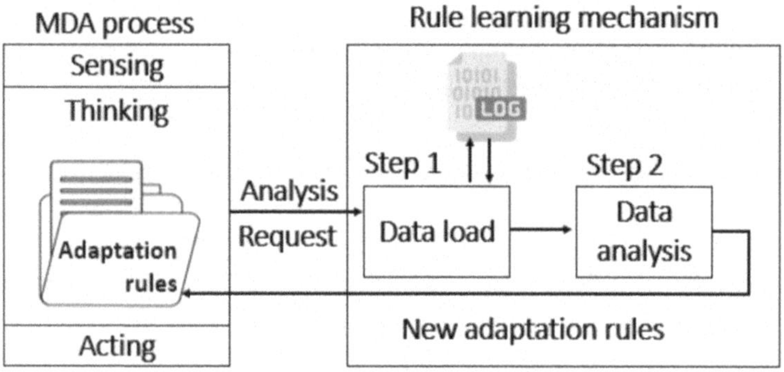

Fig. 2. Adaptation rules learning from HUD.

The sequential learning algorithm shown in the pseudo-code below is used in order to learn rules by considering one class at a time. Here, the classes refer to the adaptation actions while the features correspond to the context elements values in log data. Hence, the learning algorithm serves to classify the features (context elements values) according to each class (adaptation action), in order to build "IF – THEN" adaptation rules. For each adaptation action A_i, we search for the rules exclusively cover tuples from this action and not any other adaptation actions. The learning process includes the following operations:

- **Initialization:** it has the same role as in the basic version of sequential covering algorithm as it starts with an empty rule set.
- **Rule generation:** it selects an uncovered adaptation action from the HUD and generates a rule to cover it.
- **Rule refinement:** it iteratively refines a generated rule by adding conditions to it.
- **Rule evaluation:** it evaluates the quality of a given rule using a predefined evaluation metric. If the rule meets the quality threshold, then it is added to the adaptation rules set. Otherwise, the process continues to refine this rule.
- **Stopping criterion:** the process continues to iteratively generate, refine and evaluate each rule until a certain stopping criterion is met. Once the stopping criterion is satisfied, the rule set is considered complete.

The sequential covering algorithm allows the rules set to evolve and improve over time based on the content of HUD. This iterative process of learning and updating the rules can lead to better classification, especially in dynamic environments where contextual data change frequently over time.

Adaptation rules learning algorithm

Input: HUD as a log file (dataset)

HUD_vals, the set of all context elements (features) and adaptations actions (classes) as well as their possible values.

Output: A set of IF-THEN rules.

Begin

```
    Adaptation_rules_set = { };
    For each adaptation action A_i do
        repeat
            Rule = Learn-One-Rule (HUD, HUD_vals, A_i);
            remove tuples covered by 'Rule' form HUD;
        until (termination condition);
        Adaptation_rules_set = Adaptation_rules_set + Rule;
    End for
    Return Adaptation_rules_set;
```

End.

4 Implementation, Experiments, Results and Discussion

4.1 Implementation of the Proposed Approach

To validate our proposal, we have implemented a real prototype through which multimedia documents (web pages containing pedagogical contents) are adapted according to users context changes while storing their historical data in a log file, as shown in Table 1. The hardware configuration includes an *Arduino UNO board* endowed with *sound*, *GPS* and *photoresistor light sensors*. We have also involved an *HP Z640 station* (*Xeon* CPU with 20 cores and 64 GB of RAM) used as a server, and a *Galaxy A10s Smartphone* (*Helio P22* CPU and 2 GB of RAM) used as a client device on which the documents are run. The context values are acquired from different sources, as detailed in Table 3. Note that the quantitative context values are quantified so that they are effectively used during adaptation rules learning from log data.

Several adaptation actions are prosed to the users, in order to provide adapted contents depending on both the nature of media objects and the current context values. The description of each adaptation actions is given as follows:

- ***Speech-to-text:*** replacing auditory objects with textual content.

- ***Contrast-adjustment:*** highlighting visual objects contrast in accordance with brightness ratio.
- ***Video-to-images:*** converting videos to images sequences.
- ***Language-translation:*** translating media objects to the languages specified in users profiles.
- ***Format-change:*** changing a media object from one format to another (e.g., AVI to 3gp).
- ***Video-in-smart-TV:*** playing videos in a smart TV.
- ***Text-summarization:*** summarizing textual objects contents.

Table 3. The different context elements values.

Context element	Domain of values	Source
Noise	{*quiet, less noisy, noisy*}	Sensors
Luminosity	{*penumbra, luminous, very luminous*}	
Battery level	{*low, full*}	Device
Screen size	{*narrow, wide*}	
Smart TV	{*available, unavailable*}	
Network type	{*WI-FI, mobile data*}	
Preferred language	{*appropriate, inappropriate*}	Profile
Agenda	{*empty, busy*}	
Location	{*public place, home, laboratory, meeting room, bus*}	
Time	{*day, night*}	Device

4.2 Experiments, Results and Discussion

The following scenarios show typical cases of how HUD are used in personalizing the adaptation rules through the rule learning mechanisms discussed in Sect. 3.

> *"Ahmed is a university student. During his training, he follows online courses written in different languages. In the morning, Ahmed always takes the bus to go to the university. Along the way, he consults some online courses and therefore asks the system to convert auditory contents into text. In the afternoon, he takes a break between courses sessions in the university's public spaces. During this period, he sometimes takes a look on some online courses through his smartphone before joining the class. After getting home in the evening, Ahmed always revises his online courses while being tired. Hence, he asks the system to summarize the educational materials and to show the explanation videos in his smart TV".*

These scenarios are implemented so that the historical data of the user are stored in a log file hosted in a server. The inference of personalized adaptation rules is performed by executing the process given in Fig. 2. Table 4 shows some typical adaptation rules retained by the adopted rule-learning algorithm.

Table 4. Typical adaptation rules retained by the rule-learning algorithm.

Rules	Conflicts	Adaptation actions	Accuracy
R1	*if (network type = mobile data) then*	Format-change	93.10%
R2	*if (language = inappropriate) then*	Language-translation	100%
R3	*if (location = laboratory) then*	Speech-to-text	100%
R4	*if (time = night) then*	Summarize-text	92.85%
R5	*if (location = home and smart TV = available and screen size = narrow) then*	Video-in-smart-TV	87.33%
R6	*if (noise = noisy and location = public place and network type = mobile data) then*	Speech-to-text	94.11%

The above scenarios show how historical data can contribute to building intelligent systems by applying rule-learning algorithms to the prediction of users' behaviours, in order to infer a set of adaptation rules instead of defining them in a fixed way. More specifically, the adaptation rules represented in Table 4 only pertain to the user mentioned in the scenarios (i.e. Mr. Ahmed). In other words, a different set of adaptation rules may pertain to any other user depending on his behavior (i.e. its log data). Note that the conditions in rules depend on system usage in terms of the occurrence of context values and the decisions made by users. For instance, let consider Rules 5 and 6 in Table 4 whose conditions are inferred according to their occurrence in log data. If these rules were written by a designer, they would be as follows: *if (smart TV = available and Screen size = narrow) then Video-in-smart-TV* and *if (noise = noisy* ***or*** *location = public place) then Speech-to-text*. In fact, for Rule 5, the user uses the smart TV when he is at home and consults multimedia documents through his smartphone. Likewise, he asks the system to convert auditory contents into text when he is in a noisy public place where he plays multimedia documents by connecting to the Internet through mobile data.

More generally, this approach for building personalized adaptation rules is an unsupervised machine learning mechanism that does not need to any pre-trained data (datasets). In addition, it can easily be reused in other rule-based context-aware pervasive environments.

5 Conclusion

In this paper, a new rule-learning approach was proposed in the goal of personalizing the adaptation rules within context-aware multimedia documents adaptation processes. The proposal uses rule based data mining classifiers by means of *the sequential covering algorithm* by taking advantage of HUD resulting from the execution of these processes.

The proposal was validated by developing a real prototype that allowed us implementing two scenarios to make context-aware adaptation rules learning. As a result, the rule-learning algorithm could retain personalized adaptation rules with high accuracy degree. Note that our prototype was designed while taking into account a subset of general actions without depending on any particular application. As future work, we plan to improve the performance of rule-learning algorithms to run them on multiple devices ranging from resource-constrained devices to those with large computational capabilities.

References

1. Jannach, D., Leopold, K.: Knowledge-based multimedia adaptation for ubiquitous multimedia consumption. J. Netw. Comput. Appl. **30**(3), 958–982 (2007)
2. Laborie, S., Euzenat, J., Layaïda, N.: Semantic adaptation of multimedia documents. Multimedia Tools Appl. **55**(3), 379–398 (2011)
3. Alti, A., Laborie, S., Roose, P.: A community-based semantic social context-aware driven adaptation for multimedia documents. Int. J. Virtual Communities Soc. Networking (IJVCSN). **7**(2), 31–49 (2015)
4. Bettou, F., Boufaida, M.: An adaptation architecture dedicated to personalized management of multimedia documents. Int. J. Multimedia Data Eng. Manage. **8**(1), 21–41 (2017)
5. Saighi, A., Philippe, R., Ghoualmi, N., Laborie, S., Laboudi, Z.: Hama: a handicap-based architecture for multimedia document adaptation. Int. J. Multimedia Data Eng. Manage. **8**(3), 55–96 (2017)
6. El Guabassi, I., Bousalem, Z., Achhab, M., Al Jellouli, I., El Mohajir, B.E.: Personalized adaptive content system for context-aware ubiquitous learning. Procedia Comput. Sci. **127**, 444–453 (2018)
7. Belhadad, Y., Refoufi, A., Roose, P.: Spatial reasoning about multimedia document for a profile based adaptation. Multimedia Tools Appl. **77**(23), 30437–30474 (2018)
8. Saighi, A., Laboudi, Z., Philippe, R., Laborie, S., Ghoualmi-Zine, N.: On using multiple disabilities profiles to adapt multimedia documents: a novel graph-based method. Int. J. Inf. Technol. Web. Eng. **15**(3), 34–60 (2020)
9. Chen, G., Zhang, Y., Chen, N.S., Fan, Z.: Context-aware ubiquitous learning in science museum with iBeacon Technology. In: Spector, J.M., Lockee, B.B., Childress, M.D. (eds.) Learning, Design, and Technology. Springer, Cham (2023)
10. Zouhaier, L., BenDalyHlaoui, Y., Ayed, L.B.: Adaptive user interface based on accessibility context. Multimedia Tools Appl. **82**, 35621–35650 (2023)
11. Wongwatkit, C., Panjaburee, P.: A duplex adaptation mechanism in the personalized learning environment. J. Comput. Educ. (2023)
12. Sayed, W.S., Noeman, A.M., Abdellatif, A., et al.: AI-based adaptive personalized content presentation and exercises navigation for an effective and engaging E-learning platform. Multimedia Tools Appl. **82**, 3303–3333 (2023)
13. Bettou, F.: A new approach to processing semantic heterogeneity in adapted multimedia documents based on ontology alignment. In: Proceedings of the 9th World Congress on Electrical Engineering and Computer Systems and Sciences (EECSS 2023). Brunel University, London (2023)
14. Bettou, F., Boulkroun, B.: A multi-viewpoint approach for semantic multimedia documents adaptation. In: Proceedings of the 9th World Congress on Electrical Engineering and Computer Systems and Sciences (EECSS 2023). Brunel University, London (2023)
15. Mahalle, P.N., Dhotre, P.S.: Context-aware pervasive systems. In: Context-Aware Pervasive Systems and Applications, ch. 3, sec. 3.1, pp. 45–66. Springer, Singapore (2020)

16. Sarker, I.H.: Data science and analytics: an overview from data-driven smart computing, decision-making and applications perspective. SN Comput. Sci. **2**(5), 377 (2021)
17. Gámez, J.C., David, G., González, A., Pérez, A.: Ordinal classification based on the sequential covering strategy. Int. J. Approximate Reasoning **76**, 96–110 (2016)
18. Fürnkranz, J., Kliegr, T.: A brief overview of rule learning. In: Bassiliades, N., Gottlob, G., Sadri, F., Paschke, A., Roman, D. (eds) Rule Technologies: Foundations, Tools, and Applications. RuleML 2015. Lecture Notes in Computer Science(), vol. 9202. Springer, Cham (2015)
19. Huynh, V.Q.P., Fürnkranz, J., Beck, F.: Efficient learning of large sets of locally optimal classification rules. Mach. Learn. **112**, 571–610 (2023)
20. Rapp, M., Fürnkranz, J., Hüllermeier, E.: On the efficient implementation of classification rule learning. Adv. Data Anal. Classif. (2023)

Enhancing Efficiency and Privacy in Distributed Machine Learning: A Comparative Analysis of Federated Learning and Split Learning Techniques

Imed Eddine Bouramoul(✉), Soumia Zertal, and Makhlouf Derdour

Computer Sciences Department, LIAOA Laboratory,
University of Oum El Bouaghi, Oum El Bouaghi, Algeria
{imededdine.bouramoul,zertal.soumia,
derdour.makhlouf}@univ-oeb.dz

Abstract. As the volume of data generated by individuals and organizations continues its exponential grow, the need for efficient and secure Machine Learning (ML) algorithms has become increasingly important. Federated Learning (FL) and Split Learning (SL) are two innovative approaches designed to tackle these issues by allowing ML models to be trained using decentralized data sources while preserving data privacy. In this paper, we present a comparative analysis of FL and SL, focusing on their strengths and weaknesses in terms of accuracy, efficiency, and security. We also discuss their potential applications in various domains. Our analysis provides insights into the trade-offs between these two techniques and helps researchers and practitioners make informed decisions when choosing the appropriate approach for their specific use case.

Keywords: Federated Learning · Split Learning · Privacy · Accuracy · Comparative Analysis

1 Introduction

Distributed Machine Learning (ML) has witnessed a surge in popularity in recent years, driven by the increasing need for scalable and privacy-preserving ML solutions [1]. Federated Learning (FL) and Split Learning (SL) have become leading techniques in the field of distributed ML, enabling multiple clients to collaboratively train ML models without revealing their original data. These methods have found applications across diverse domains such as healthcare, finance, and the Internet of Things (IoT). Despite their respective merits and drawbacks, it is crucial to conduct a comparative analysis of FL and SL to discern their strengths and limitations in various contexts.

This paper provides a comparative analysis of FL and SL, focusing on their technical aspects, performance, and privacy. Specifically, we compare FL and SL in terms of privacy and security, communication efficiency, computation efficiency, accuracy,

M. R. Laouar et al. (Eds.): ICISAT 2023, LNNS 981, pp. 224–232, 2024.
https://doi.org/10.1007/978-3-031-60591-8_19

scalability, and applicability to different scenarios. We also discuss the challenges and opportunities of FL and SL in distributed ML.

Following this introduction, the succeeding segments of this paper are organized in the following manner: In Sect. 2, we present an in-depth examination of FL and SL, encompassing their fundamental principles, and real-world applications. Section 3 undertakes a comprehensive comparative analysis of FL and SL, scrutinizing aspects like privacy and security, communication efficiency, computational efficiency, scalability, and adaptability to diverse scenarios. Section 4 delves into a discussion of the challenges and prospects inherent to FL and SL within the domain of distributed ML. Lastly, in Sect. 5, we draw the paper to a conclusion, while also outlining potential avenues for future research in the realm of FL and SL.

2 Federated Learning and Split Learning Overview

2.1 Federated Learning

Federated Learning (FL) represents a ML approach that empowers multiple participants to collaboratively train a model while preserving the confidentiality of their individual data. The field of privacy-preserving ML has gained significant prominence in practical applications, sparking heightened interest from both academic and industrial sectors. FL boasts promising applications across various domains, including healthcare, finance, and social media, and distinguishes itself by allowing data to stay on the devices or nodes where it was initially collected, rather than undergoing central server transfer. This approach safeguards sensitive data from exposure to third parties [1].

FL reinforces privacy and security by employing a decentralized data model, implementing local training on each device, which implies that data remains confined to the device and remains inaccessible to others. This practice mitigates the risk of data breaches and unauthorized access [2]. Moreover, FL integrates differential privacy techniques to introduce noise into the data, rendering individual data points challenging to identify. This dual mechanism protects data privacy while still facilitating effective model training [3].

Homomorphic encryption stands as another critical element in FL, encrypting data before transmission to the central server. This measure ensures data remains private and secure, even if intercepted by unauthorized parties [4]. Lastly, FL employs secure aggregation techniques [5] to merge local models into a unified global model, preserving data privacy and security throughout the aggregation process.

2.2 Advantages of Federated Learning

FL is a ML method that empowers organizations to train Artificial Intelligence (AI) models using decentralized data, eliminating the requirement to centralize or disclose that data. The state of the art presented in [6, 7] and [8] has allowed us to summarize them as follows: firstly, FL ensures data privacy by keeping data decentralized and local, furthermore, it permits multiple users to collaboratively build a centralized ML model without the necessity of divulging private user data; secondly, FL diminishes communication expenses as data remains distributed and doesn't necessitate transfer to a central location. This reduces the need for high-bandwidth communication channels and reduces the risk of data breaches, additionally, FL can be expanded to encompass millions of devices and supports retaining complete control over data. This allows organizations to uphold ownership and management of their data while still facilitating its use in ML. This makes it easier to train models on large datasets without the need for a centralized server; finally, FL allows for the creation of a single model that's been trained using multiple datasets, without any of those datasets being shared (see Fig. 1). This stands in contrast to traditional ML approaches where all the datasets are centralized and uploaded to one server.

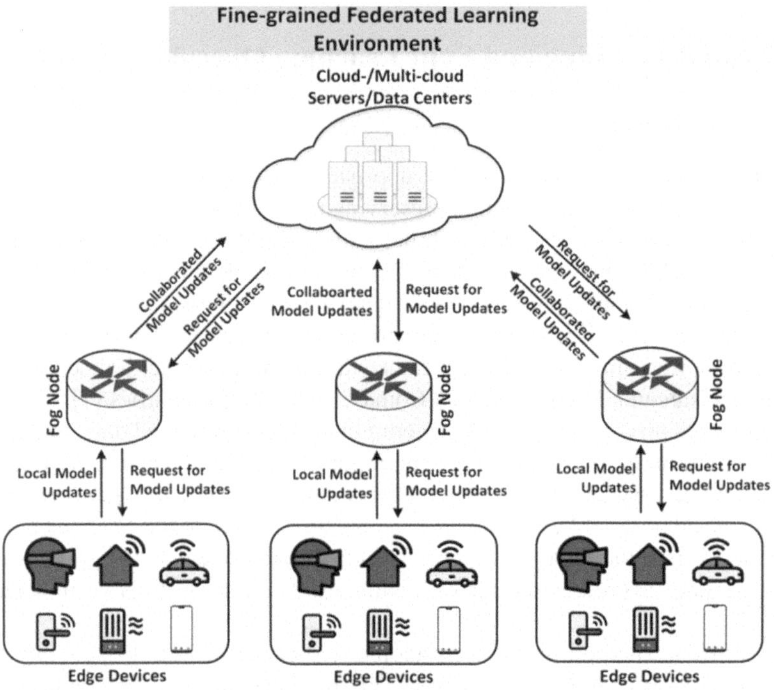

Fig. 1. FL architecture overview [9].

2.3 Federated Learning Uses Case

FL has been successfully deployed in various real-world applications, including virtual keyboards, IoT applications, health care, and smart healthcare applications. Recent research has evaluated data-centric challenges in FL, including attributes like data partitioning traits, data distributions, data security measures, and benchmark datasets.

2.4 Split Learning

SL is a distributed ML technique that allows multiple entities to train Deep Neural Networks (DNN) without sharing raw data or detailed information about the model [10]. In SL, a DNN undergoes division into multiple segments, each of which receives training on distinct clients. The data undergoing training remains confined to the respective client's device, with only the output of the partial model shared with other entities [11]. SL preserves data privacy by enabling clients to collaboratively train a Deep Learning (DL) model with the server without the necessity of sharing any raw data. The server receives intermediate representations or processed data, and the training process is finalized without accessing the original data [12]. SL provides an intuitive level of data privacy, and it is similar to FL, but it differs in that FL requires all participating clients to upload their model updates to the server, while SL only requires the client to send the output of the cut layer to the server [13].

2.5 Advantages of Split Learning

SL has several advantages over traditional ML approaches, including reduced communication costs, improved privacy, and scalability. SL can be employed to train DNN using data from multiple sources without the necessity of directly sharing the raw labeled data [14]. A study of works presented in [15, 16] and [17] has allowed us to summarize advantages of SL as follows: SL reinforces model confidentiality by separating the model from the server (see Fig. 2). This makes SL more privacy-preserving than FL, and SL has lower communication costs compared to FL.

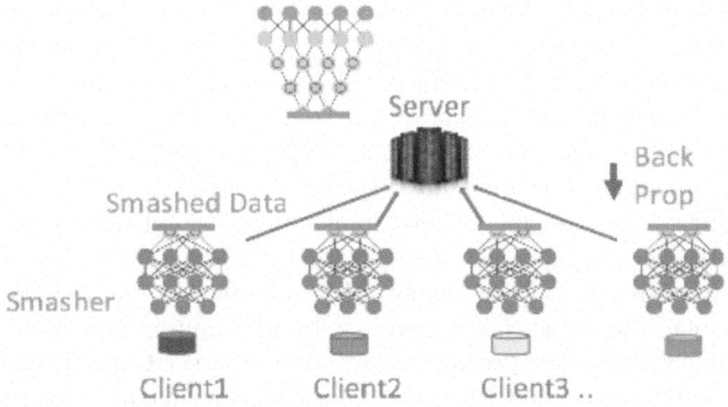

Fig. 2. SL architecture overview [17].

2.6 Split Learning Uses Case

SL has several advantages over FL in specific use cases, including healthcare, smart cities, industrial IoT, resource-constrained devices, and improved privacy. However, both approaches have different use cases and can be applied to different scenarios depending on the specific requirements of the application.

3 Comparative Analysis Federated Learning and Split Learning

Most existing works have compared FL and SL on one or two aspects at most, such as privacy or scalability. This makes it difficult to choose between these two techniques. Our analysis will take into consideration several comparison aspects such as privacy, scalability, computation, communication load, and other parameters. This will help researchers make rational decisions when choosing between FL and SL.

a) Comparison of data privacy mechanisms: FL and SL each come with their distinct strengths and limitations when it comes to safeguarding data privacy. FL enables several entities (clients) to collaborate in solving a ML problem without sharing their raw data with each other or with a central server [20]. SL, on the other hand, splits a DNN into multiple sections, each of which is trained on a different client. Moreover, both approaches have unique advantages and limitations in terms of data privacy. In SL, the intermediate representations or smashed data are transmitted over the wire, which provides an intuitive level of data privacy. However, it is difficult to quantify how much protection the method provides. SL can also be combined with homomorphic encryption to train models on encrypted data [19]. FL, on the other hand, provides a quantifiable measure of data anonymization through differential privacy [21]. Differential privacy protection makes it possible to measure the privacy loss when a model is trained on a dataset, and it can be used to ensure that the model does not memorize sensitive user data. FL can also be integrated with additional privacy-preserving techniques, including secure multi-party computation and homomorphic encryption;
b) Comparative Analysis of Communication Overhead: The primary distinction in communication overhead between FL and SL lies in the manner in which the ML model is partitioned between server and clients. In FL, the server's training tasks are broken down and distributed to clients for parallel learning, which amplifies the computational load on clients and results in increased communication overhead for model exchanges between the server and clients [22]. Conversely, SL divides the ML model into a client and a server model at a specified cut layer, with devices dedicated to training their respective models and transmitting the cut layer's activations to the server [23];
c) Comparative Analysis of Computational Requirements: As per [24], in FL, clients play a pivotal role in the training process by transmitting individual local updates, which originate from their private data, to a central server. Consequently, FL demands higher computational resources on the client side compared to SL, as clients are tasked with executing local model updates and computations. On the other hand, the server's resource demands in FL can be less intensive than in SL since it only needs to manage a single model instead of maintaining separate models for each client.

   Within SL, the ML model is divided into two segments. Devices are solely responsible for training their designated model and forwarding the activations of the cut layer to the server. Additionally, only a small fraction of the model is allocated to the clients for training, with the bulk of the model retained on the server, consequently alleviating the computational burden on the client side in SL. However, the model partitioning strategy used in SL introduces a notable increase in communication costs, as it necessitates the transmission of activations from the cut layer from clients to the server. Ultimately, on the server side, SL may impose more substantial resource requirements compared to FL, as it necessitates maintaining separate models for each client;

d) Comparative Model Size Analysis: The dimensions of the model can impact communication costs within FL as a consequence of the necessity to transmit model parameters between clients and the central server. This transmission becomes particularly costly when resource-limited clients collectively endeavor to train a substantial ML model, as they contribute by conveying local updates derived from their private data to the coordinating server [6]. On the other hand, SL provides an inherent solution in these situations where only a portion of the model is stored and trained on clients, while the majority of the model remains solely on the servers. However, the model partitioning approach used in SL does result in a significant increase in communication costs;

e) Comparative Analysis of Client Numbers: According to [25], it is evident that an increasing number of clients tilts the balance in favor of the SL approach. In simpler terms, as the number of clients grows, SL becomes more efficient in terms of communication. However, the model partitioning method used in SL does introduce a significant rise in communication costs when compared to the traditional FL approach;

f) Comparative Analysis of Accuracy: An in-depth comparative analysis of FL and SL in [17] highlighted the fact that FL experiences a greater decrease in accuracy than SplitFed as the proportion of malicious clients increases. In addition, a recent study [15] compared SplitFed, a hybrid of FL and SL, and the pure SL and FL approaches in terms of test accuracy and communication efficiency. The results indicate that SplitFed achieves similar test accuracy and communication efficiency to SL, while significantly reducing communication costs compared to the pure FL approach.

4 Challenges and Opportunities

Despite the advantages brought by the two approaches, several challenges are faced and research work is needed for the overcome. The challenges turn around communication overhead, data heterogeneity, and privacy/security. FL requires communication between clients and server, which can be a bottleneck in terms of bandwidth, latency, and energy consumption. Furthermore, FL operates under the assumption that the data is distributed in an Identically and Independently Distributed (IID) manner among the clients, which might not hold true in real-world scenarios. Non-IID data can result in suboptimal model performance and convergence issues [26]. FL involves sharing data and model updates between clients and server, which can raise privacy and security concerns. Differential privacy and secure aggregation have been proposed to address these issues [27]. On the contrary SL introduces a significant amount of communication cost, as the activations of the cut layer need to be transmitted from the client to the server. The method of splitting the model in SL introduces a potential risk of data leakage, as it might enable the server to reconstruct input data by leveraging the correlation between input and intermediate activations [23]. Finally, SL assumes that the data is IID across the clients, which may not be true in practice. Non-IID data can lead to poor model performance and convergence

[27]. To overcome the challenges already presented, the searchers offer approaches that combine the advantages of FL and SP. Various techniques have been proposed to address scalability issues in FL and SL, including Communication and Storage Efficient Federated and Split Learning (CSE-FSL), FedLite, Hybrid Federated Split Learning (HSFL) [18], split FL based on client side clustering, and Byzantine-resilient distributed gradient descent algorithm. These techniques aim to reduce communication and computation costs while maintaining high learning accuracy and data privacy.

5 Conclusion

To summarize, FL and SL are notable distributed collaborative learning techniques that prioritize data privacy by enabling clients to retain control of their private data without the need to share it with other clients or servers. Each of these techniques possesses its unique strengths and weaknesses concerning accuracy, scalability, efficiency, and security. FL grapples with issues like communication costs, constraints associated with edge device computing capabilities, and privacy concerns. Conversely, SL demands the thoughtful design of model architecture and effective management of communication costs. Our comparative analysis concentrated on evaluating accuracy, efficiency, and security aspects in both FL and SL.

It was observed that the choice of the layer at which the model is divided between clients and server can influence the effectiveness of security measures. FL experiences a more pronounced drop in accuracy as the proportion of malicious clients increases when compared to SplitFed. Despite the challenges they present, FL and SL have found extensive applications in diverse domains like smart healthcare, smart cities, and smart industry. Researchers have proposed various techniques, including compression algorithms and efficient communication strategies, to address some of these challenges. Nevertheless, continuous research and advancements are essential to tackle these challenges and enhance the overall performance of FL and SL. The evolution of distributed collaborative learning methods, such as FL and SL, contributes significantly to the understanding and advancement of ML and its myriad applications across various fields.

References

1. Mao, J., Cao, C., Wang, L., Ye, J., Zhong, W.: Research on the security technology of federated learning privacy preserving. J. Phys. Conf. Ser. **1757**, 012192 (2021). https://doi.org/10.1088/1742-6596/1757/1/012192
2. Asif, H.M., Karim, M.A., Kausar, F.: Federated learning and its applications for security and communication. Int. J. Adv. Comput. Sci. Appl. **13** (2022). https://doi.org/10.14569/IJACSA.2022.0130838
3. Ma, C., et al.: On safeguarding privacy and security in the framework of federated learning. IEEE Netw. **34**, 242–248 (2020). https://doi.org/10.1109/MNET.001.1900506
4. Bai, Y., Fan, M.: A method to ımprove the privacy and security for federated learning. In: 2021 IEEE 6th International Conference on Computer and Communication Systems (ICCCS), pp. 704–708. IEEE, Chengdu, China (2021). https://doi.org/10.1109/ICCCS52626.2021.9449214

5. Nguyen, T., Thai, M.T.: Preserving privacy and security in federated learning. IEEE/ACM Trans. Netw. 1–11 (2023). https://doi.org/10.1109/TNET.2023.3302016
6. Mu, Y., Shen, C.: Communication and Storage Efficient Federated Split Learning. http://arxiv.org/abs/2302.05599 (2023)
7. Federated Learning. https://rapidminer.com/glossary/federated-learning/, Accessed 04 Nov 2023
8. Vepakomma, P., Gupta, O., Swedish, T., Raskar, R.: Split learning for health: distributed deep learning without sharing raw patient data. http://arxiv.org/abs/1812.00564 (2018)
9. Ur Rehman, M.H., Salah, K., Damiani, E., Svetinovic, D.: Towards blockchain-based reputation-aware federated learning. In: IEEE INFOCOM 2020 - IEEE Conference on Computer Communications Workshops (INFOCOM WKSHPS), pp. 183–188. IEEE, Toronto (2020). https://doi.org/10.1109/INFOCOMWKSHPS50562.2020.9163027
10. Vepakomma, P., Swedish, T., Raskar, R., Gupta, O., Dubey, A.: No Peek: A Survey of private distributed deep learning. http://arxiv.org/abs/1812.03288 (2018)
11. Vepakomma, P., Raskar, R.: Split learning: a resource efficient model and data parallel approach for distributed deep learning. In: Federated Learning: A Comprehensive Overview of Methods and Applications, pp. 439–451. Springer International Publishing, Cham (2022). https://doi.org/10.1007/978-3-030-96896-0_19
12. Duan, Q., Hu, S., Deng, R., Lu, Z.: Combined federated and split learning in edge computing for ubiquitous intelligence in internet of things: state-of-the-art and future directions. Sensors. **22**, 5983 (2022). https://doi.org/10.3390/s22165983
13. Zhang, Z., Pinto, A., Turina, V., Esposito, F., Matta, I.: Privacy and Efficiency of Communications in Federated Split Learning. http://arxiv.org/abs/2301.01824 (2023)
14. Ayad, A., Renner, M., Schmeink, A.: Improving the communication and computation efficiency of split learning for IoT applications. In: 2021 IEEE Global Communications Conference (GLOBECOM), pp. 01–06. IEEE, Madrid (2021). https://doi.org/10.1109/GLOBECOM46510.2021.9685493
15. Thapa, C., Chamikara, M.A.P., Camtepe, S., Sun, L.: SplitFed: When Federated Learning Meets Split Learning. http://arxiv.org/abs/2004.12088 (2022)
16. Khan, M.A., Shejwalkar, V., Houmansadr, A., Anwar, F.M.: Security Analysis of SplitFed Learning. http://arxiv.org/abs/2212.01716 (2022)
17. Singh, A., Vepakomma, P., Gupta, O., Raskar, R.: Detailed comparison of communication efficiency of split learning and federated learning. http://arxiv.org/abs/1909.09145 (2019)
18. Cheng, Z., Fan, X., Liwang, M., Min, M., Wang, X., Du, X.: Hybrid Architectures for Distributed Machine Learning in Heterogeneous Wireless Networks. http://arxiv.org/abs/2206.01906 (2022)
19. Khan, T., Nguyen, K., Michalas, A.: Split Ways: Privacy-Preserving Training of Encrypted Data Using Split Learning. http://arxiv.org/abs/2301.08778 (2023)
20. Abuadbba, S., et al.: Can We Use Split Learning on 1D CNN Models for Privacy Preserving Training?. http://arxiv.org/abs/2003.12365 (2020)
21. Lin, H., Liu, W., Wang, X.: A secure federated learning mechanism for data privacy protection. In: 2021 20th International Conference on Ubiquitous Computing and Communications (IUCC/CIT/DSCI/SmartCNS), pp. 25–31. IEEE, London (2021). https://doi.org/10.1109/IUCC-CIT-DSCI-SmartCNS55181.2021.00019
22. Duan, J., Duan, J., Wan, X., Li, Y.: Efficient federated learning method for cloud-edge network communication. In: 2023 5th International Conference on Communications, Information System and Computer Engineering (CISCE), pp. 118–121. IEEE, Guangzhou (2023). https://doi.org/10.1109/CISCE58541.2023.10142819
23. Kim, M., DeRieux, A., Saad, W.: A Bargaining Game for Personalized, Energy Efficient Split Learning over Wireless Networks. http://arxiv.org/abs/2212.06107 (2022)

24. Wang, J., Qi, H., Rawat, A.S., Reddi, S., Waghmare, S., Yu, F.X., Joshi, G.: FedLite: A Scalable Approach for Federated Learning on Resource-constrained Clients. http://arxiv.org/abs/2201.11865 (2022)
25. Gao, Y., et al.: End-to-End Evaluation of Federated Learning and Split Learning for Internet of Things. http://arxiv.org/abs/2003.13376 (2020)
26. Khan, L.U., Saad, W., Han, Z., Hossain, E., Hong, C.S.: Federated Learning for Internet of Things: Recent Advances, Taxonomy, and Open Challenges. http://arxiv.org/abs/2009.13012 (2021)
27. Abdelmoniem, A.M., Ho, C.-Y., Papageorgiou, P., Bilal, M., Canini, M.: On the Impact of Device and Behavioral Heterogeneity in Federated Learning. http://arxiv.org/abs/2102.07500 (2021)

A Model of Collaborative Filtering Based on Trust Communities in Complex Networks

Abdelhani Diboune[1,2](✉), Hachem Slimani[3], Kadda Beghdad Bey[4], and Hassina Nacer[1]

[1] MOVEP Laboratory, USTHB, BP 32, 16111 Bab Ezzouar, Algiers, Algeria
adiboune@usthb.dz

[2] ATE Department, Faculty of Technology, University of Bejaia, 06000 Bejaia, Algeria

[3] LIMED Laboratory, Faculty of Exact Sciences, University of Bejaia, 06000 Bejaia, Algeria
hachem.slimani@univ-bejaia.dz

[4] Informatics Systems Laboratory, Ecole Militaire Polytechnique, BP 17, 16046 Bordj El Bahri, Algiers, Algeria

Abstract. In this paper, we propose a model of Social-based collaborative filtering that considers the reciprocal influence of users trust communites. This model consists mainly of three main steps: In Step 1, social communities in the trust social network are detected using a Genetic algorithm, and the items' similarity communities are computed. In Step 2, based on the depicted community structures and the rating data, a probabilistic model is developed to predict efficient recommendations. Finally, in Step 3, given a targeted user and a set of potential items, the learned probabilistic model is used to predict his/her preferential in order to propose him/her relevant recommendations. On the other hand, an experimental study was conducted to evaluate the cohesiveness of trust communities and the general accuracy of recommendations on *Rich Epinion Dataset*. The obtained results have shown that the proposed model outperforms some traditional methods of collaborative filtering.

Keywords: Social networks · community detection · social collaborative filtering · genetic algorithms

1 Introduction

The proliferation of information networks has improved the accessibility to online information. However, this richness brought a new challenge: the information overload. Indeed, users are overwhelmed by large amount of useless information. In this scope, recommender systems can be a good solution to deal with this issue by suggesting to users only the items and services which best fit their tastes and interests.

Recommender systems aim to estimate the preference of a user for new items and services in the network based on their historical information inputs such as explicit ranking or implicit feedback, for instance, browsing history, research patterns and so forth. In this context, recommender systems use a plethora of methods from many fields such

M. R. Laouar et al. (Eds.): ICISAT 2023, LNNS 981, pp. 233–243, 2024.
https://doi.org/10.1007/978-3-031-60591-8_20

as information retrieval, data mining and machine learning. These Recommender systems may be classified into three main categories: *Collaborative filtering*, *Content-based approaches*, and *Hybrid approaches* which combine the benefits of both approaches [1]. Collaborative filtering [2] is based on the similarity computation to predict ratings. The main idea consists of seeking for similarities between the active user and the other users in the system and predict his/her ratings for new items, according to ratings of similar users. Another principle consists in finding similar items clusters for each item in the system and the recommendation is made based on the neighborhood of items preferred by the active user. On the other hand, in Content-based recommender system [3], each user is assigned a profile which describes his/her preferences. This description consists of a set of features which characterize purchased items. The recommendation in such systems implies the comparison of extracted features from new items with descriptions in the active user profile. Thus, items similar to the user profile are recommended to him/her.

Despite numerous academic efforts on the subject, recommender systems continue to face a number of challenges. First, there is an urgent need to improve the prediction accuracy of recommender systems on a consistent basis. Second, the data sparsity and cold-start issues afflict collaborative filtering performance. The sparsity problem stems from the fact that the user/item rating matrix is quite sparse. In fact, despite the large number of products, in recommender systems, each user typically rates only a handful. To deal with these issues, *Social recommender systems* use the social network features. They are mainly based on the fact that users are likely to trust their friend's recommendations more than similar unconnected users. Hence, such systems examine the targeted user social/trust network and ascertains his/her friend's preference over the given item. Then, it generates recommendations for highly rated items.

In this context, we propose in this paper a model of Social trust-based collaborative filtering based on evolutionary community detection. It mainly consists of three steps: In the first step, trust communities in the social network are detected using a Genetic algorithm. Furthermore, items' similarity communities are extracted using *K Nearest Neighbors* (KNN) algorithm. In the second step, based on rating data and the community structures detected in the previous step, a probabilistic model is developed to predict accurate recommendations of items to users. Finally, in the third step, given a targeted user and a set of items, the learned probabilistic model is used to predict his/her preferential such that relevant recommendations are proposed to him/her.

The reminder of this paper is organized as follows: The next section is a synthesis of related works approaches. In Sect. 3, we describe our model of collaborative filtering based on trust communities. Section 4 details experimental results. Finally, Sect. 5 concludes the paper and cites some relevant perspectives.

2 Related Works

Works in literature which are most related to our paper fall into two main headings: *Collaborative filtering* and *Collaborative filtering based on social networks*. In this section, we give a synopsis of the relevant works for each class.

2.1 Collaborative Filtering

Collaborative filtering intends to predict the preference of a user to particular items based on his/her previous likings and the opinion of other like-minded users in the system. In a collaborative filtering, there is a set of n users $U = \{u_1, u_2, \ldots, u_n\}$, a set of m items $I = \{i_1, \ldots, i_m\}$, and for each user $u_i \in U$ there is a subset $I_{u_i} \subset I$ of items which the user u_i has rated. Given an active user $u_a \in U$, the aims of collaborative filtering is to find items with highest utility $argmax_{i_k \in I} R(u_a, i_k)$ to the user, where $R(u_a, i_k)$ is the rating of the user u_a of the item i_k. Broadly speaking, we distinguish two different kinds of collaborative filtering methods: *Memory based* and *Model based collaborative filtering*.

2.1.1 Memory Based Collaborative Filtering

Memory based collaborative filtering algorithms use the entire user-item rating matrix to predict utilities of items to the active user. The first step consists of depicting in the systems subsets of users with similar interests (neighbors). Then, the prediction of preferences is computed by aggregating the ratings of neighbors. In literature, such techniques are known as *Nearest neighbors collaborative filtering* [4]. Memory based collaborative filtering are lazy learning algorithms, and induces many shortcoming such as low performance in the case of space rating matrix. Further more, this class of recommender systems suffers from the cold start problem, meaning that these algorithms find difficulties to recommend for new users in the network.

2.1.2 Model Based Collaborative Filtering

A global model is learned to fit the training data. Then, the recommendations for test datasets or real world queries are predicted based on this model. In this scope, there is a large volume of published studies describing the role of many models such as *Singular Value Decomposition* (SVD), *Latent semantic indexing* (LSI), *Principal Component Analysis* (PCA), *Association Retrieval Technique* [5], etc. on learning Collaborative filtering models to predict real world rating data and overcoming the drawbacks of memory based methods such as data sparsity and scalability. However, the main drawback of these approaches is that they are less accurate than the memory based approaches [6].

2.2 Collaborative Filtering Based on Social Networks

It is now well established that memory based and model based collaborative filtering performances are limited by data sparcity and scalability issues of recommender system [7]. In this scope, Shi et al. [8] proposed an approach based on *Latent factor analysis* using *Probabilistic matrix factorisation*. The main idea consists of learning users and items latent feature spaces by employing both social network and user-item matrix. This method deals with data sparsity and poor prediction accuracy problems. Furthermore, Li et al. [9] computed every rating $r_{i,j}$ in the user-item matrix by aggregating both of the user taste and weighted mean of the predicted ratings for item j from the user u_i trusted-friends preferences. In order to learn a recommendation model, this method uses a *Linear combination of the matrix factorization* approach to predict ratings. As this method is based on *Eager learning*, depicting a model on training data is offline. Thus, processing queries are quicker. However, the main inconvenient is that Matrix Factorization in whole rating matrix still time-consuming.

3 The Proposed Collaborative Filtering Model Based on Trust Communities

In this section, we describe the proposed social collaborative filtering model based mainly on community detection for users and items. In fact, as shown in Fig. 1, the model consists of three main steps: *Community detection for users and items*, *Learning probabilistic model*, and *Recommendation phase*. In what follows, we describe the different steps.

Step 1 (Community detection for users and items): This step consists of two main parallelizable tasks: *Trust community detection of users* and *Similarity community detection of items.*

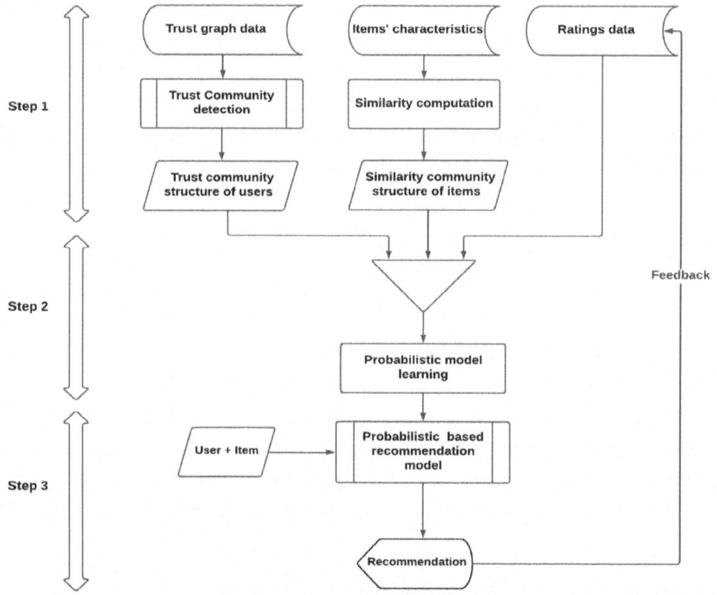

Fig. 1. Collaborative filtering model based on trust communities

(i). **Trust community detection of users:** This step entails locating cohesive trust circles in graph databases. Community detection can be modeled as a Multi-objective optimization problem. The social network is represented as a graph $G(V(G), E(G))$, in which $V(G)$ is the vertex set representing users, $E(G) \subset V(G)^2$ is edge set that represents social interactions or relationships (trust, fellowship, friendship, etc.). The objective is to cluster the network into modules (subnetworks) with high inter-cluster density and low intra-cluster density.

In this context, we chose the two fitness functions which are based on two metrics: $L(C_i, C_i) = \sum_{v,w \in C_i} A_{v,w}$ and $L(C_i, \overline{C_i}) = \sum_{v \in C_i,\, w \notin C_i} A_{v,w}$, such that C_i is a community in the graph dataset, $\overline{C_i} = V(G) \setminus C_i$, and A is the adjacency matrix of the interconnection graph.

- The first fitness function called *Negative Ratio Association* (NRA) is defined by the formula: $NRA = -\sum_{i=1}^{k} \frac{L(C_i, C_i)}{|C_i|}$. To minimize this objective, nodes within the cluster $C_i, \forall i \in \{1, \ldots, k\}$, should be densely connected.
- The second fitness is *Ratio Cut* (RC) which is defined by the formula: $RC = \sum_{i=1}^{k} \frac{L(C_i, \overline{C_i})}{|C_i|}$. To minimize this objective, the cluster $C_i, \forall i \in \{1, \ldots, k\}$, should be loosely connected to the rest of the network.

To solve this problem, we opted for optimization by *Genetic algorithms* where the main operations are presented as follows:

Population initialization: In our approach, the generation of the initial population is not arbitrary, we use instead *Random walk initialization.* Firstly, a node in the network is chosen uniformly and considered as the starting node of the random walker. Then, an outgoing edge is chosen randomly and moves to the next node. From that node, we choose a neighbor at random and move to it, etc. This sequence of nodes forms a random walk on the interconnection graph. The nodes belonging to the same random path are affected to the same initial community. After the execution of a random walk, another unaffected node is chosen randomly, and the random walk algorithm is executed in the subgraph in which nodes are not affected yet to a community. Indeed, the clusters in graphs are generally persistent structures with respect to some random walk after a given number of steps t. Which means that the probability that the walker escapes from that cluster is very low. As result, nodes within the same random walk have higher probability to be effectively in the same community. Thus, this initialization allows a quicker convergence of the Genetic algorithm.

Crossover: For a best exploration of the individual space, our approach uses *Standard uniform crossover*. This method uses a binary mask Msk of length equal to the number of nodes which is generated randomly. An offspring X is generated from two parents X_1, X_2 by selecting gens from the first parent for which the corresponding mask value is 0, and from the second parent for which the mask value is 1.

$$X[i] = \begin{cases} X_1[i], \ \textit{if } Msk[i] = 0, \\ X_2[i], \ \textit{if } Msk[i] = 1. \end{cases}$$

Mutation: A node i is chosen at random, then is affected randomly to another cluster among its neighbors.

Selection: At every iteration, the individuals are ranked according to the number of individuals dominated by them. The choose of individuals for the next iteration in the Genetic algorithm uses roulette wheel algorithm over this fitness: $p_i = \frac{f_i}{\sum_i f_i}$, where f_i is the number of solutions which are *Pareto-dominated* by the individual X_i.

(ii). **Similarity community detection of items:** This task consists on creating an *Ego-centred community network* for each item $i \in I$, where I the whole set of items, by calculating the similarity between items. In this context, our method defines the community of an item $i \in I$ as the set $CI(i)$ of *K Nearest Neighbors* (KNN), *i.e.* most similar items.

Step 2 (Learning probabilistic model): To decide if an item $i^{(k)} \in I, i^{(k)} = \left(i_1^{(k)}, \ldots, i_m^{(k)}\right)^T$ may be recommended to a user$u^{(l)} = \left(u_1^{(l)}, \ldots, u_n^{(l)}\right)^T$, the model estimates the probabilities of ratings of the user $u^{(l)}$ on the item$i^{(k)}$, *i.e.*$R\left(u^{(l)}, i^{(k)}\right) = a$, such that $a \in \{1, \ldots, 5\}$. Let $CU(u^{(l)})$ be the trust community of the user $u^{(l)}$ that has been computed in the trust community detection step (Step 1(i)), and let $CI(i^{(k)})$ be the cluster of items similar to the item $i^{(k)}$ that has been depicted previously in the item community detection step (Step 1(ii)).

$$P_a = P\left[R(u^{(l)}, i^{(k)}) = a|u^{(l)}, i^{(k)}, R(u^{(s)}, i^{(t)}) = a\right], s.t.u^{(s)} \in CU\left(u^{(l)}\right) \wedge i^{(t)} \in CI(i^{(k)}).$$

$$P_a = P\left[R\left(u^{(l)}, i^{(k)}\right) = a|u^{(l)}, i^{(k)}\right] \times P\left[R\left(u^{(l)}, i^{(k)}\right) = a|R(u^{(s)}, i^{(t)}) = a)\right], s.t.u^{(s)} \in CU\left(u^{(l)}\right) \wedge i^{(t)} \in CI\left(i^{(k)}\right),$$

where $P\left[R\left(u^{(l)}, i^{(k)}\right) = a|u^{(l)}, i^{(k)}\right]$ is the probability that the user $u^{(l)} \in U$ gives to the item $i^{(k)} \in I$ a rating $R(u^{(l)}, i^{(k)}) = a$ given the item's characteristics $i^{(k)} = \left(i_1^{(k)}, \ldots, i_m^{(k)}\right)^T$ and user's attributes$u^{(l)} = \left(u_1^{\{(l)\}}, \ldots, u_n^{(l)}\right)^T$. This probability represents the predicted compatibility between the users' taste and the item based on the previous ratings of the user. This value is learned by a *Multinomial logistic regression* as follows:

For each couple $(u^{(l)}, i^{(k)}) \in U \times I$, we consider a row wise concatenation vector. $X^{(l,k)}$ defined by:

$$X_i^{(l,k)} = \begin{cases} 1, \; if\; j = 0, \\ u_j^{(l)}, \; if\; j \in \{1, \ldots, n\}, \\ i_{(j-n)}^{(k)}, \; elsewhere. \end{cases}$$

For each user $u^{(l)} \in U$, $IR\left(u^{(l)}\right)$ is the set of items rated by the user $u^{(l)}$.

The existing ratings in the system are considered as a training dataset$DSet = \{(x^{(l,k)}, R(u^{(l)}, i^{(k)})), \forall u^{(l)} \in U, i^{(k)} \in IR(u^{(l)})\}$.

$$\forall u^{(l)} \in U, \forall i^{(k)} \in IR\left(u^{(l)}\right), \forall a \in \{1, \ldots, 5\} : P\left[R\left(u^{(l)}, i^{(k)}\right) = a|u^{(l)}, i^{(k)}\right] = h_\Theta\left(X^{(l,k)}\right)$$
$$= \frac{1}{1 + e^{\Theta^T X^{(l,k)}}},$$

where $\Theta = (\Theta_0, \ldots, \Theta_{k+l})^T$ is a vector of parameters to be learned in order to minimize the following cost function:

$$J_a(\Theta) = \frac{1}{|DSet|} \sum_{(X^{(i,k)}, R(u^{(l)}, i^{(k)})) \in DSet} y_a^{(l,k)} \log\left(h_\Theta\left(X^{(l,k)}\right)\right)$$
$$+ \left(1 - y_a^{(l,k)}\right) \log\left(1 - h_\Theta\left(X^{(l,k)}\right)\right),$$

where $y_a^{(l,k)} = \begin{cases} 1, ifR\left(u^{(l)}, i^{(k)}\right) = a, \\ 0, elsewhere. \end{cases}$

In this context, the *Limited memory Broyden Fletcher Goldfarb Shanno* algorithm (L-BFGS) is accurate to solve the optimization problem in an acceptable computing time.

On the other hand, the second factor of the P_a formula: $P\left(R\left(u^{(l)}, i^{(k)}\right) = a \middle| R\left(u^{(s)}, i^{(t)}\right) = a, \text{ where } u^{(s)} \in CU\left(u^{(l)}\right) \wedge i^{(t)} \in CI(i^{(k)})\right)$ is the probability that the user $u^{(l)} \in U$ gives to item $i^{(k)}$ a rating $R(u^{(l)}, i^{(k)}) = a$ given the ratings assigned by users belonging to the trust community of $u^{(l)}$ to the items similar to$i^{(k)}$. As consequence, it represents the influence of the trust communities over the user choices.

$$P\left[R\left(u^{(l)}, i^{(k)}\right) = a \middle| R\left(u^{(s)}, i^{(t)}\right) = a\right] = \frac{\left|U_{v \in CU\left(u^{(l)}\right)}\left\{R\left(v, i^{(k)}\right) = a \middle| i^{(k)} \in IR(v)\right\}\right|}{\left|U_{v \in CU\left(u^{(l)}\right)}\{IR(v)\}\right|},$$

$$s.t. u^{(s)} \in CU\left(u^{(l)}\right) \wedge i^{(t)} \in CI\left(i^{(t)}\right).$$

Step 3 (Recommendation phase): This step consists on recommending potential item $i^{(k)}$ to a targeted user $u^{(l)}$. Indeed, once the Probabilistic based recommendation model is learned, the probability that the user $u^{(i)}$ attributes a rating $R(u^{(l)}, i^{(k)}) = a$ to the item $i^{(k)}$ is computed for every $a \in \{1, \ldots, 5\}$. Then, the rating value $R_p(u^{(l)}, i^{(k)})$ is chosen as follows: $R_p\left(u^{(l)}, i^{(k)}\right) = \underset{\{1,\ldots,5\}}{\text{argmax}} P_a$. Finally, if $a \geq 3$, then the item $i^{(k)}$ is recommended to the user $u^{(l)}$.

4 Results of Experiments and Analysis

In this section, we conduct experimental studies on *Rich Epinion Dataset* (RED) to evaluate the performance of our approach. We assess both the performance of our Genetic algorithm based Trust community detection model and the accuracy of the Probabilistic based recommendations. Before presenting the simulations results, we describe the used dataset and software development environment.

4.1 Dataset

We use *Rich Epinions Dataset* (RED) as a Recommender system dataset to conduct our experiments [10]. The dataset contains 131 228 users, 317 755 items and 1 127 673 reviews, that is a 0.003% sparsity. UML class diagram of RED dataset has the following tables: *User:* contains the attributes name, location, top rank and profile visits count. *Item:* contains name, category and profile url. *Category:* contains name, parent category, description url, lineage (path in the category tree) and depth (in the category tree). *Review:* it associates a user with an item and contains the rating between 1 and 5, the review rating (the mean of all review ratings associated with this review), and the review date. *Expertise:* users who are considered as experts in some categories are associated with the corresponded to these categories. *Similarity:* the similarity between all users couples are computed using the Pearson coefficient correlation.

4.2 Software Development Environment

The simulation was conducted using *Python* programming language with necessary packages in it. Namely, *NumPy* package for representing multidimensional array objects such as tensors, arrays, matrices, etc. as well as implementing fast operations on them, *Pytorch* to implement optimization algorithms such as L-BFGS, *NetworkX* to provide different functionalities of network representation and implement some graph clustering algorithms.

4.3 Evaluation of Community Detection Accuracy

The quality of the partitioning is evaluated by using the *Normalized Mutual Information* (NMI). Let *A* and *B* be two partitioning of a network represented by the graph *G(V,E)*, let *C* be the confusion matrix, such that $C_{i,j}$ is the number of nodes of the community *i* of the partition *A* that are also in the community *j* of the partition *B*. Thus, the formula of NMI is defined as follows:

$$NMI(A,B) = \frac{\sum_{i=1}^{N_A}\sum_{j=1}^{N_B} C_{ij}\log\left(C_{ij}n/\left(C_{i.}C_{.j}\right)\right)}{\sum_{i=1}^{N_A} C_{i.}\log\left(C_{i.}/n\right)\sum_{j=1}^{N_B} C_{j.}\log\left(C_{j.}/n\right)},$$

where N_A and N_B are respectively the number of communities in the partition *A* and *B*, $n = |V|$ is the number of nodes, $C_{i.} = \sum_j C_{ij}$ and $C_{.j} = \sum_i C_{ij}$.

The NMI values obtained from the simulation of our proposed Genetic algorithm based community detection model over the trust network of RED dataset are presented in Fig. 2.

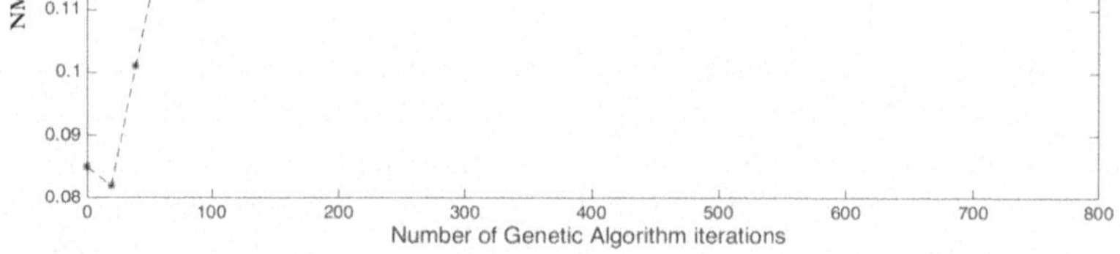

Fig. 2. Normalized mutual information community detection algorithm.

Figure 2 clearly shows a good performance of the Genetic algorithm based community detection model on the trust network of the RED dataset which is a very large network. Indeed, the method produces an NMI value of approximatively of 0.16, after the 700th iteration, which can be considered as a very acceptable accuracy considering the size and the sparsity of the studied trust network.

4.4 Evaluation of Recommendation Accuracy

To evaluate the performance of the recommender algorithm, we use *F-Measure*, which is computed as follows: First, we extract the values of *True Positives* (TP), *True Negatives* (TN), *False Positives* (FP), and *False Negatives* (FN) from the *Confusion matrix*. Then, *Precision* and *Recall* are computed according to the following equations: $Precision = \frac{TP}{TP+FP}$, $Recall = \frac{TP}{TP+FN}$.

Finally, *F-Measure* is computed using the following equation:

$$FMeasure = 2 \times \frac{Precision \times Recall}{Precision + Recall},$$

The simulation results are described in Table 1.

From Table 1, the conclusions that stand out can be summarized in the following points:

- In general, our model establishes a good accuracy of recommendations. Indeed, it generally achieves an F-Measure higher than 0.6, which confirms the importance of joining the trust data to the recommender system.
- The results revealed a correlation between the number of communities and the accuracy. Indeed, the model shows different accuracy performances for predicting recommendations. This is mainly due to the fact that the granularity of community structure affects the quality of graph trust clustering. More specifically, according to the size of communities, the depicted coalitions could have different purchasing correlations (adding irrelevant users or removing relevant ones).
- By using our proposed genetic algorithm based community detection approach to extract trust communities, the recommendation shows a significant enhancement over the two other approaches based on Newman and Girvan algorithm [11], and Lancichinentti et al. algorithm [12], which is mainly due to the quality of the used fitness functions. Indeed, instead of using one objective function that quantifies both inter-cluster density and inter-cluster sparsity, we have used a multi-objective optimization approach which induces a better clustering results.
- Using Ego-centred communities performs the worst results in terms of accuracy, this can be explained by the information loss induced by neglecting non-neighboring nodes in trust cluster.

Table 1. Comparative table of F-Measures obtained by our approach using different numbers of trust communities and some concurrent approaches in the literature.

Approaches used in Trust communities detection		F-measure
Ego Centred Network Based		0.601
Girvan and Newman Based [11]		0.652
Lancichinetti et al. Based [12]		0.751
Our approach	Number of communities = 5	0.62
	Number of communities = 12	**0.79**
	Number of communities = 15	0.74
	Number of communities = 17	**0.76**
	Number of communities = 20	**0.82**

5 Conclusion

In this paper, we propose a trust community-based social collaborative filtering approach which combines both recommendations and genetic algorithms based community detection approach. In this method, through using the users' trust social relations networks, the trust communities are detected for the network users. Once the community to which a targeted user belongs is identified, based on the users' historical interests belonging to that community, items are added to the candidate recommendation list according to a probabilistic model. After predicting the rate of target user for each candidate item, using the ratings of the same items attributed by the trusted users, the items whose ratings are greater than a given threshold are recommended to the targeted user. The results of simulations on the RED dataset reveal that the proposed method outperforms some existing methods in literature. Furthermore, by proceeding in this way the problem of scalability in data is mitigated by the formation of community structure. Also, the issue of cold start problem is handled since the proposed method uses trust relations.

As further perspectives of our research work, it would be interesting to assess the effects of using overlapping trust communities in our proposed recommender system. Furthermore, in our approach, the influences of different individuals among the trust community are assumed to be equal. Thus, further study is needed to be undertaken considering the expertise attribute of each user. Indeed, the influence of a user should depend on his/her expertise level of each category of items. Finally, further experimental studies need to be carried out in order to validate the performance of our approach compared with other relevant recent proposals in the context of recommender system.

References

1. Gediminas, A., Tuzhilin, A.: Toward the next generation of recommender systems: a survey of the state-of-the-art and possible extensions. IEEE Trans. Knowl. Data Eng. **17**(6), 734–749 (2005)

2. Koren, Y., Steffen, R., Robert, B.: Advances in collaborative filtering. In: Recommender Systems Handbook, pp. 91–142 (2021)
3. Pazzani, M.: Content based recommendation systems [in:] the adaptive web: methods and strategies of web personalization. In: Brusilovsky, P., Kobsa, A., Nejdl, W. (eds.) Springer Verlag, Berlin Heidelberg (2007)
4. Balabanović, M., Shoham, Y.: Fab: content-based, collaborative recommendation. Commun. ACM **40**(3), 66–72 (1997)
5. Yang, X., Guo, Y., Liu, Y., Steck, H.: A survey of collaborative filtering based social recommender systems. Comput. Commun. **41**, 1–10 (2014)
6. Su, X., Khoshgoftaar, T.M.: A survey of collaborative filtering techniques. Adv. Artif. Intell. **2009**, 421425 (2009)
7. He, J., Chu. W.W.: A social network-based recommender system (SNRS). In: Data Mining for Social Network Data, pp. 47–74. Boston, MA: Springer US (2010)
8. Shi, W., Lu, Y., Huang, Z., Du, E., Cheng, Y., Chen, J.: Personalized recommendation leveraging social relationship. In: The 8th International Conference on Computer Engineering and Networks (CENet2018). Springer International Publishing (2020)
9. Li, M., Huang, D., Wei, B., Wang, C.D.: Event recommendation via collective matrix factorization with event-user neighborhood. In: Intelligence Science and Big Data Engineering: 7th International Conference, IScIDE 2017, Dalian, China, September 22–23, 2017, Proceedings 6, pp. 676–686. Springer International Publishing (2017)
10. Meyffret, S., Guillot, E., Médini, L., Laforest, F.: RED: a Rich Epinions Dataset for Recommender Systems. [Research Report] LIRIS (2012)
11. Newman, M.E., Girvan, M.: Finding and evaluating community structure in networks. Phys. Rev. **69**(2), 026113 (2004)
12. Lancichinetti, A., Fortunato, S., Kertész, J.: Detecting the overlapping and hierarchical community structure in complex networks. New J. Phys. **11**(3), 033015 (2009)

Author Index

M. R. Laouar et al. (Eds.): ICISAT 2023, LNNS 981, pp. 245–246, 2024.
https://doi.org/10.1007/978-3-031-60591-8